Cosmic **Horizons**

Cosmic **Horizons**

Astronomy at the Cutting Edge

Steven Soter and Neil deGrasse Tyson, Editors

AN AMERICAN MUSEUM OF NATURAL HISTORY BOOK

The New Press, New York

Cover: The spiral galaxy NGC 4414 has a dense central bulge composed of old red stars, surrounded by spiral arms of young blue stars and lanes of absorbing interstellar dust. The galaxy is 56,000 light-years across and 2 million light-years away.

Right: The Sagittarius star cloud lies toward the center of our Milky Way Galaxy. Hubble Space Telescope.

Published in the United States by The New Press, New York, 2001
Distributed by W. W. Norton & Company, Inc., New York

LIBRARY OF CONGRESS CATALOGING-IN-PUBLICATION DATA
Cosmic Horizons: Astronomy at the Cutting Edge/Steven Soter
and Neil deGrasse Tyson, editors.
p. cm.
ISBN 1-56584-602-8 (pbk.)
1. Astronomy. I. Soter, Steven. II. Tyson, Neil deGrasse.
QB51. C735 2001
520—dc2100–051101

The New Press was established in 1990 as a not-for-profit alternative
to the large, commercial publishing houses currently dominating
the book publishing industry. The New Press operates in the public
interest rather than for private gain, and is committed to publishing,
in innovative ways, works of educational, cultural, and community
value that are often deemed insufficiently profitable.

The New Press, New York
450 West 41st Street, 6th floor
New York, NY 10036

www.thenewpress.com

Printed in England

987654321

Contents

Section One: Solar Systems

Section Two: Stars

The Cygnus Loop is an expanding blast wave from
a star that exploded about 15,000 years ago.
Hubble Space Telescope

Section Three: Galaxies

Section Four: Universe

Foreword Ellen V. Futter

Since its founding in 1869, the American Museum of Natural History has put the world on display. The twin pillars of our mission have always been science and education. Today, the Museum is one of the world's leading research institutions in the natural sciences. Over the years, our scientists—now a staff of more than 200 men and women—have gone on more than 100 expeditions a year. They collect evidence from all over the globe in their effort to answer questions about such fundamental scientific and human issues as the origins of the universe, Earth and life, who we are and where we fit. In addition to research, our scientists have a related responsibility—to interpret science for the general public. The exhibitions at the Museum have been conceived and curated by scientists who are committed to putting the evidence— "the real stuff"—in front of the public.

As we move into the twenty-first century, we at the Museum are filled with a renewed dedication to our mission. To many people, science today seems too removed and too difficult to understand, yet the need has never been greater for a public that is well informed about science. Through its educational initiatives and exhibitions, the Museum seeks to narrow the gap between what people know and what they need to know about science. To that end, in 1997 the Museum launched the National Center for Science Literacy, Education, and Technology to extend the Museum's resources beyond its walls to a national audience.

Science is exploration. Scientists work on the frontier—on the border of the known and the unknown. This book series, through the words of working scientists, enables non-scientists to share the excitement of cutting-edge science— the excitement of discovery. The series includes four volumes that expand the themes covered in many of our major new exhibitions.

Our exhibitions always embody a scientist's vision and point of view. In the same way, each book in this series is "curated"—researched, organized, and introduced—by one of the Museum's scientists. Each book features a selection of essays written by leading scientists who have made significant contributions to the field. The essays are supported by case studies and profiles of important people and events.

This volume, *Cosmic Horizons: Astronomy at the Cutting Edge*, explores the mysteries and wonders of the universe to expand the themes addressed in our new Rose Center for Earth and Space and Cullman Hall of the Universe. Other volumes in this series include: *Epidemic!: The World of Infectious Disease*, which explores the themes of the major special exhibition at the Museum in 1999 by the same name; *The Biodiversity Crisis: Losing What Counts*, which focuses on biodiversity, and the interrelatedness of all living things on our planet to expand the themes addressed in the Hall of Biodiversity; and *Earth: Inside and Out,* which focuses on the Earth, and the phenomena that shape our planet to expand the themes addressed in the Gottesman Hall of Planet Earth.

This series illustrates our continuing commitment to connect the general public with the natural world. We cannot send real specimens to every home and classroom, but we can bring the ideas, concerns, and questions of working scientists directly to you. We hope these books provide a valuable resource that will prepare tomorrow's leaders to make informed decisions about the world and the universe we all share.

Ellen V. Futter is President of the American Museum of Natural History.

Acknowledgments

This book was produced by the National Center for Science Literacy, Education, and Technology, American Museum of Natural History.

Ellen V. Futter, President

Myles Gordon, Vice-President for Education

Nancy Hechinger, Director of the National Center for Science Literacy, Education, and Technology

The National Center would like to acknowledge the National Aeronautics and Space Administration for its general programmatic support and for the support for this book series.

Editorial Staff:
Senior Writer and Editor: Steven Soter
Associate Editor: Ellen Przybyla
Contributing Writers:
Benjamin R. Oppenheimer and Ashton Applewhite

Production Staff:
Project Director: Caroline Nobel
Creative Director: Patricia Abt
Production Manager: Ellen Przybyla
Production Coordinator: Eric Hamilton
Production Assistant: Ethan Davidson

Design by Sheena Calvert, parlour, NY

We would like to thank Diane Wachtell, Ellen Reeves, Barbara Chuang, Gary Tooth, Leda Scheintaub, and Fran Forte at The New Press.

Exterior of the American Museum of Natural History

By any measure,
we are living in a golden
age of astronomy.

The moon Triton beyond the horizon of Neptune.
Voyager 2 Spacecraft.

Preface: On the Cosmic Horizon

Neil deGrasse Tyson

By any measure, we are living in a golden age of astronomy. The total number of astrophysicists engaged in research is fifty percent more than just a decade ago. The unprecedented number of large telescopes on Earth and in orbit has provided more access to the universe than ever before. And the rate at which research papers are being published is the highest ever.

But before we go too far congratulating ourselves on our remarkable productivity, consider that when a field of research grows exponentially, every generation thinks of itself as living in a special time. For the *Astrophysical Journal*, a leading research publication in the field, half of the last century's published papers appeared in the past fifteen years. With true exponential growth, astrophysicists could have made the identical statement fifteen years ago. As they could fifteen years before that. And fifteen years before that. And fifteen years before that.

Therein lies the challenge to science museums. How do you design a facility that teaches the basics of the universe but also tracks its cutting edge? In conceiving and designing the Frederick Phineas and Sandra Priest Rose Center for Earth and Space, the most ambitious project ever undertaken by the American Museum of Natural History, we identified multiple ways to bring the universe to the public. For example, when we display planetary orbits around the Sun, or the basic tenets of stellar evolution, the exhibit evokes a feeling of security in our knowledge. But when we display the latest theories on the formation of planets, stars, galaxies, and the universe, we present them in the form of video interviews with leading researchers in the field, so that we can update

them with new discoveries when they occur. For this collection of essays, conceived at the same time as the exhibits for the Rose Center, we have unabashedly stepped onto this volatile frontier by assembling the contributions of two dozen astrophysicists. They supply for the reader a snapshot of this exponential growth in our cosmic knowledge at the beginning of the twenty-first century. We make no attempt to recreate Astro 101. Rather, we offer the reader a hand-picked sampling of the breadth of observational, theoretical, and technological developments in modern astrophysics.

Not long ago, the community of astronomers comprised just two camps: the observers, who used telescopes, and the theorists, who used pencil and paper. Today, the growth of technology, above all else, drives the advances in astrophysics. For example, discoveries of planetary science flow principally from data beamed back to Earth from dozens of space probes, which have transformed our solar system into a veritable experimental laboratory for comparative planetology and for exobiology. And there is now a new breed of scientist, the computational astrophysicist, whose tools are the laws of physics and whose workspace is the computer code that probes the behavior of the universe in otherwise untestable ways. For example, the growing speed of computers has so radically transformed the study of interstellar clouds (where most stars are born) that nearly all of what we know about the detailed behavior of such clouds comes to us not from direct observation but from these laboratories of cyberspace.

I cannot imagine what a few more decades will bring. If the exponential growth of discovery continues to hold, then the early twenty-first

Neil deGrasse Tyson is the Frederick P. Rose Director of New York City's Hayden Planetarium and a member of the Department of Astrophysics at the American Museum of Natural History.

century will look like primitive times, where research moved like molasses. Our Golden Age of astronomy flows not from the uniqueness of a prodigious rate of research, but from the consistency with which we have maintained this exponential growth. By that measure, our Golden Age has been running strong since the Industrial Revolution, with no sign of letting up. And there will be no rest for editors of volumes such as this as we try to stay abreast of the cosmic frontier, just as our brethren did hundreds of years before.

We have organized *Cosmic Horizons* into six sections, each of which resonates strongly with prevailing themes of the Rose Center. The first four sections sort the contents of the universe according to the principal areas relevant to astrophysics: Section One: Solar Systems; Section Two: Stars; Section Three: Galaxies; and Section Four: Universe. The editors take pleasure in naming Section One in the plural, which reflects the explosive growth in the discovery of other planetary systems. These four sections are followed by Section Five: Life in the Universe and Section Six: Technical Frontiers. For Section Five, we have openly recognized the enormous public interest in the question, "Are we alone in the universe?" And no discussion of modern astrophysics would be complete without Section Six, a testimonial to the continued role of technology in pushing back the frontier as fast and as far as possible.

Our invited authors share with the reader their knowledge of what is current, their passion for what is hot, and their vision for what may soon alter how we view our place in the universe. Each of their essays is meant to stand alone. Interspersed among them, we have provided profiles and historical case studies about astronomers whose seminal work transformed the field in their own day. We have also supplemented the volume with an extensive glossary of scientific terms and concepts that appear in these readings.

Our goal, from the beginning of this project, has been to empower the reader to peer into the intersection of what is known and unknown in the universe. We do so while not forgetting that as the area of our knowledge grows, so too does the perimeter of our ignorance.

Messier 80, one of the densest globular clusters in our Galaxy, is a swarm of several hundred thousand stars about 28,000 light-years away. Hubble Space Telescope.

Montage of Saturn and its moons. Voyager Spacecraft.

Section One: Solar Systems

Introduction Steven Soter

As astronomers explore the geology and meteorology of the planets, they also devote increasing attention to the smaller worlds of our solar system—the diverse collection of moons, asteroids, and comets. At the same time, other worlds are being discovered in orbits around distant suns. This section samples some important advances in the study of the smaller bodies of our solar system and in the discovery of other solar systems.

Our solar system includes four small, rocky worlds—Mercury, Venus, Earth, and Mars—in orbits rather close to the Sun. Farther out lie four giant planets—Jupiter, Saturn, Uranus, and Neptune—all made of gases and ices surrounding rocky cores. These gas giants have thick atmospheres, orbiting rings, and families of icy satellites. Between the inner terrestrial and the outer gas giant planets is a belt of tens of thousands of rocky asteroids, the source of most meteorites. At the outer edge of our planetary system is the Kuiper Belt, a broad band of comets orbiting the Sun. Extending halfway to the nearest star, lies the spherical Oort cloud, source of all the long-period comets that occasionally plunge to the inner solar system.

The planets all orbit the Sun in the same direction and in nearly the same plane. The Sun itself slowly spins in the same sense. This shared angular motion of the Sun, the planets, and the Kuiper Belt objects is a remnant of the solar nebula, the rotating disk of gas and dust from which the solar system formed about 4.6 billion years ago. The more distant Oort cloud comets shows no preference for the orbital plane or direction of the planets, but can move in any direction.

Among the dozens of moons of the solar system, the "the big seven" stand out as substantial worlds in their own right—the Earth's Moon, the four Galilean moons of Jupiter (Io, Europa, Ganymede, and Callisto), Titan of Saturn, and Triton of Neptune. At more than 2,700 kilometers in diameter, they are all larger than Pluto. Two of them, Ganymede and Titan, are actually larger than the planet Mercury, and Titan has an atmosphere denser than the Earth's.

Our Moon is a dry, rocky hulk, which makes it exceptional among the larger satellites. It was evidently too hot when it formed to acquire any water or other volatiles (substances that melt or vaporize at relatively low temperatures). In striking contrast, the giant satellites circling the outer planets formed far enough from the Sun to acquire volatiles in abundance. In the cold outer solar system, one might expect these satellites to be frozen solid. But some of them are heated by friction from periodic tidal deformation of their solid interiors. This internal energy supply is intense enough to drive astonishing geological activity on these satellites. One such world is Europa, the smallest of the four Galilean satellites of Jupiter. It shows every sign of having a deep ocean of liquid water under an icy crust. And where there's liquid water and flowing energy, there may be life.

Steven Soter is a member of the Department of Astrophysics at the American Museum of Natural History in New York.

The solar system formed from a rotating disk of gas and dust. The central concentration became the Sun, while the material in the disk agglomerated to form the planets. The energetic young Sun expelled jets of gas from its poles.

The solar system continues to produce surprises. In 1992, David Jewitt and Jane Luu discovered the first member of an extensive family of comets in a broad belt with its inner edge near the orbit of Neptune. This new feature of the solar system was named the Kuiper Belt, after the astronomer Gerard Kuiper, who predicted its existence in 1951. The Kuiper Belt is much nearer to the Sun than the Oort cloud. About a third of the known Kuiper Belt objects share the same motion as Pluto, which crosses the orbit of Neptune and revolves around the Sun twice during every three revolutions of Neptune. To understand the overall structure of our solar system, it makes more sense to think of Pluto as the largest known Kuiper Belt Object than as a planet. But some astronomers disagree with this view.

The current controversy over how to classify Pluto has a close historical precedent. In 1801, Giuseppi Piazzi discovered the first asteroid, Ceres, in an orbit between Mars and Jupiter. Astronomers at once assumed it was a new planet, but many similar objects were soon found in the same part of the solar system. These objects were then recognized as members of a class distinct from the planets, and came to be known as asteroids. Similarly, when Clyde Tombaugh discovered Pluto in 1930, everyone welcomed it as the ninth planet. More than sixty years later, when many more icy objects were found in similar orbits, astronomers recognized that they had discovered another new class of solar system objects, which constitute the Kuiper Belt.

So what is a planet? It may be hard to believe, but at the moment there is no general agreement on a simple definition. We propose the following working definition for a planet: "A celestial object with enough mass for its gravity to make it spherical but not enough to trigger any nuclear reactions. Unlike satellites, planets orbit stars in non-intersecting orbits or drift freely in space." Pluto and Ceres are massive enough to be spherical, but their orbits intersect those of other members of their class, so they are not planets by the above definition. Objects larger than planets and smaller than stars are called brown dwarfs; they have a feeble early stage of nuclear fusion, in contrast to stars, which have sustained intense nuclear fusion. Our definition would accommodate so-called interstellar "rogue planets," which are not tied to any star. Astronomers theorize that many such objects may exist in interstellar space, but such unilluminated bodies would be difficult to detect.

Some astronomers prefer to define the term planet based not on a body's intrinsic properties (mass and orbital characteristics) but on the mode of its formation (see the essay by David Black). But the intrinsic properties of a substellar mass are observable while its origin generally is not. For example, an interstellar body of low mass might have accreted in orbit around a star before being expelled, or it might have formed independently in space. An intrinsic definition would unambiguously categorize it as a rogue planet, a brown dwarf or a star, depending on its mass.

Objects with planetary mass have recently been discovered in orbit around other stars. Most astronomers consider these objects to be planets. But in nearly all of the systems, the "giant planets" are much closer to the star than they are in our solar system. Perhaps this is partly due to the observational selection effects that make it especially difficult to detect systems like our own. There is every reason to think that planets are forming within rotating disks observed around very young stars (see the essays by C. Robert O'Dell and by Steven V. W. Beckwith in Section Two). But some of the companions of stars actually detected so far may have been formed like brown dwarfs or stars, directly from the breakup of interstellar clouds, and without the intermediary state of a rotating circumstellar disk. More observations should determine whether or not these newly-found objects were formed like the planets in our solar system.

Earth was formed by the collision of small bodies in the solar nebula. Some of the debris left over from the formation of the solar system is still falling on our planet. Every few years, a house-sized cometary or asteroidal fragment plunges into the Earth's atmosphere, delivering a burst of energy comparable to that of the Hiroshima bomb. We don't notice it only because our deep and thick atmosphere provides an effective shield to disintegrate such relatively small and fragile objects. But over longer periods of time, much larger objects arrive. Kilometer-sized impactors simply shove the atmosphere aside, depositing most of their enormous energy on Earth's surface. One of them wiped out the dinosaurs and most other species on our planet 65 million years ago. The probability of such a catastrophic impact in our lifetime is extremely low, but not negligible. It reminds us that the Sun and Moon are not the only parts of the solar system that profoundly affect life on Earth.

To explore our solar system and others, we pose the following questions:

Does an ocean hide beneath the icy crust of Jupiter's moon Europa? If so, might it support life?

Clark R. Chapman, an Institute Scientist at the Southwest Research Institute in Boulder, Colorado, examines the strange surface of Europa for clues about what may lie below.

Why does the Kuiper Belt interest both solar system and stellar astronomers?

Jane Luu, Associate Professor of Astronomy at the Leiden Universitry in The Netherlands, co-discovered the Kuiper Belt. Here she describes the Belt and the evidence it may preserve regarding the origin of our solar system and others.

How have astronomers detected unseen bodies orbiting other stars?

David C. Black, Director of the Lunar and Planetary Institute and Vice President for the University Space Research Association Space Programs, explains the methods astronomers use for detecting the invisible companions of stars.

What are Near Earth Objects? How often do they hit the Earth?

David Morrison, Director of Astrobiology and Space Research at NASA's Ames Research Center, describes the surveys to identify these objects and determine if the Earth is at risk of a catastrophic collision anytime in the foreseeable future.

18 | 19

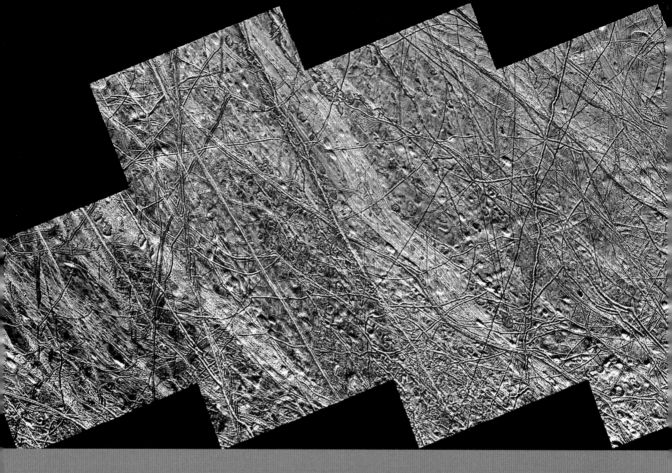

A Hidden Ocean on Europa?

Clark R. Chapman

While we often hear about the search for life on Mars, there is another, perhaps even more promising candidate for life in the solar system—Europa, the second of Jupiter's four big moons. Until two decades ago, we assumed that the moons of the outer solar system must be lifeless. They were simply too far outside the so-called "habitable zone" (in which our own planet orbits), where the Sun's heat allows the vital substance water to exist in liquid form (see the essay by Christopher Chyba in Section Five). But planetary exploration has taught us new ways to look at the universe. It now seems likely that the

Clark R. Chapman is an Institute Scientist at the Department of Space Studies of the Southwest Research Institute in Boulder, Colorado.

Mosaic image from NASA's Galileo spacecraft shows a ridged plain on Europa's northern hemisphere. Double ridges crossing the scene could be remnants of water or slushy ice eruptions that quickly froze on the cold surface. The origin of their brown color is unknown. The underlying (older) blue surface is almost pure water ice. The image covers an area about 800 by 350 kilometers (500 by 220 miles).

ingredients necessary for the origin of life existed on Europa and that a nurturing environment may still remain, just beneath the little world's icy crust.

When I was a teenager studying Jupiter and its moons with my backyard telescope, I learned almost everything that professional astronomers then knew about Europa. It was gleaming white in color and smaller than the other big moons of Jupiter—the innermost Io and the more distant Ganymede and Callisto. It was even possible to calculate the average density of Europa, using the available crude measurements of its diameter and estimates of its mass (based on mutual gravitational interactions among the satellites). Europa's density suggested a relatively "rocky" composition. Ganymede and Callisto had lower densities, suggesting more "icy" compositions.

In the early 1970s, astronomers measured the infrared spectrum of sunlight reflected from Europa's surface, which revealed the telltale absorption features of frozen water. It was no surprise that rocky Europa was covered with ice. Astronomers had already shown that Saturn's rings were made of ice, and water was known to be a cosmically abundant molecule. Also, everyone knew that it must be extremely cold at Jupiter's distance, where sunlight is only four percent as intense as on Earth.

But then another source of heat was discovered in the Jupiter system. Jupiter's powerful gravity raises strong tides in the solid bodies of its big moons. Shortly before the Voyager 1 spacecraft arrived at Jupiter in 1979, Stanton Peale and two colleagues had calculated that gravitational interactions between Jupiter's

moons prevent their orbits from becoming circular. As a result, the tidal deformation of Io varies with the satellite's changing orbital distance from Jupiter. This periodic flexing of Io produces tidal friction in the satellite, greatly raising its internal temperature. The theory predicted that enough heat would be dissipated within Io to cause observable volcanic eruptions. This was soon confirmed when the camera of Voyager 1 revealed sulfurous volcanic plumes spewing from the surface of Io.

When the tidal heating theory was then extended to Europa, it raised a fascinating possibility: even though Europa is farther than Io from Jupiter's powerful tidal influence, its interior might still be heated enough to maintain an ocean of liquid water beneath the solid icy surface.

NASA had already begun to build the Galileo spacecraft, which was intended to orbit Jupiter to explore the giant planet's atmosphere and extensive magnetosphere and, with lower priority, to look at its four big moons. But following Galileo's launch in 1989, the spacecraft's main antenna failed to open. With decreased communications capability, the plan to monitor Jupiter's cloud motions with movies became impractical. Also, after the spacecraft went into orbit around Jupiter in 1996, intense radiation from the planet kept it from approaching Io too closely. So the focus shifted to Europa.

What Galileo found on Europa was pregnant with implications. So fascinating were some of the close-up pictures of Europa's surface that NASA, despite its stretched budget, was persuaded to extend Galileo's two-year orbital mission into the year 2000.

Even from the distant passes of the two Voyager flybys, Europa had appeared enigmatic. Its generally snow-white surface was crisscrossed by a pattern of lines. Most were straight, but some followed a scalloped course. Europa's surface also exhibited some mottled, circular, and wedge-shaped darker patches. There were few obvious impact craters.

But Galileo flew much closer to Europa, and its camera resolved details more than 100 times finer than those seen by Voyager, revealing features as small as a house. These stunning views transformed Europa's rather bland appearance into a wealth of intricate topographic patterns. While few features stand more than a couple hundred meters above the average level, much of the small world's surface is covered with superimposed icy patterns of bewildering complexity (Figure 1). Double ridges extend hundreds of kilometers across the landscape, cutting through older ridged terrains (Figure 2). Long wedge-shaped features appear to be places where vast stretches of ice cracked open along fault lines and spread apart, to be filled by darker-colored ices welling up from below.

Other fascinating landforms, termed "lenticulae," were revealed within the mottled terrains. These generally circular features, typically five to twelve kilometers across,

Figure 2: Close-up of Europa's ridged plains, from NASA's Galileo spacecraft, indicate recent geological activity. The plains contain many parallel and cross-cutting ridges, often in pairs, apparently due to upwelling material from below. Valleys between some of the ridges contain dark material of unknown origin. This image is about twenty kilometers (twelve miles) across. Sunlight is coming from the upper left, casting shadows from the ridges towards the right.

were hard to interpret when observed indistinctly during Galileo's first orbit of Jupiter. They resembled eroded impact craters, like those seen on most other moons. If so, they seemed to suggest an old surface.

The geologist Eugene Shoemaker had developed the science of dating surfaces in the solar system based on counting impact craters. He calculated the expected impact rate based on a census of comets and asteroids capable of striking a particular body. He could then determine the time required to accumulate the observed number of craters—the geological age of the surface. Shoemaker initially concluded that Europa's surface was geologically old, perhaps a billion years. After all, any place that has accumulated numerous impact craters must have been around for a long time.

Subsequent closer flybys of Europa by Galileo soon clarified the evidence for craters: there were hardly any. A few large, circular, dark spots turned out to be craters or scars formed when a projectile actually punctured the ice layer, leaving concentric fractures like the result of a bullet piercing a window. Almost all of the smaller spots were revealed to be various kinds of lenticulae, which look like blisters, cavities, and collapsed ice bubbles, some encircled by moat-like depressions, formed by some kind of bubbling-up process from below. Europa's surface preserves few impact craters from asteroids and comets. This is remarkable because the Jupiter system is located in a cosmic shooting gallery, as witnessed by the large number of impact craters on two other moons of Jupiter and by the 1994 crash of Comet Shoemaker-Levy 9 into Jupiter itself.

Given the number of Jupiter-zone comets discovered by Shoemaker and others, the rarity of primary impact craters on Europa means that most of its surface must be only about 30 million years old. In geological terms, this is amazingly young, less than one percent of the age of the solar system. Among the solid-surfaced planets, satellites, and asteroids that have been examined so far, only Io and the tectonically active regions on Earth are demonstrably younger. The discovery of Europa's youthful surface by the Galileo spacecraft shows once again that the outer solar system is far from being geologically dead.

On February 20, 1997, during its sixth orbit of Jupiter, the Galileo camera finally produced dramatic digital images of Europa that, for the first time, resembled something familiar (Figure 3).

To Arctic explorer Max Coon, the pictures looked exactly like ice rafts floating in a refrozen ocean. Everything else on Europa had been like a tapestry of geometrical, modern-art designs. Now we had pictures whose interpretation seemed obvious at a glance: blocks of rigid crust had broken off and floated in a once liquid sea, twisting and tipping, until the sea once again froze in place. From the minimal numbers of impact craters on this province—named Conamara Chaos—the oceanic activity must have occurred within the last few million years.

The discovery made front-page news: "Icy Find on Jupiter's Moon may be a Sign of Ocean Life." The oceanographer John Delaney, who uses submersible vehicles like Alvin to explore the Earth's deep ocean floors, told the press that he now believed there was life on Europa. But not so fast! Although life as we know it requires liquid water, liquid water may occur without life. Furthermore, although most experts think that liquid water exists beneath the surface of Europa today, the evidence is still indirect.

If the satellite is geologically active, then liquid water should occasionally break through to the surface. But Galileo's camera did not find plumes of boiling and evaporating mist. On the other hand, we haven't closely observed much of Europa for very long. Even on our own geologically active planet, volcanic explosions may last only a few hours, separated by decades of inactivity. Perhaps in a place where Galileo's cameras never pointed there's a region of current activity on Europa, which briefly erupts before the loss of heat to frigid space once again freezes over any exposed body of water. Indeed, the form of Europa's scalloped ridges implies that the cracks responsible for them had propagated across hundreds of kilometers in just a few days or weeks. Years or centuries of stability may separate such episodes of activity. The ice rafts in Conamara Chaos may well have floated

during a relatively brief active episode in the geologically recent past.

At 3,140 kilometers in diameter, Europa is a little smaller than Earth's Moon. During numerous flybys, ground tracking of Galileo's radio signal permitted an accurate assessment

Figure 3: The Conamara Chaos region on Europa, as seen by NASA's Galileo spacecraft, shows relatively recent resurfacing of the satellite. Large ice floes, formed by the breakup of an icy crust, have shifted and rotated; some have been tipped and partly submerged. The underlying material is thought to be either liquid water or an icy slush. Younger fractures cut through the region, showing that the surface was refrozen into solid ice. The image shows an area about thirty-five by fifty kilometers (twenty by thirty miles). Sunlight is from the right. Topographical features with higher surface relief cast longer shadows to the left.

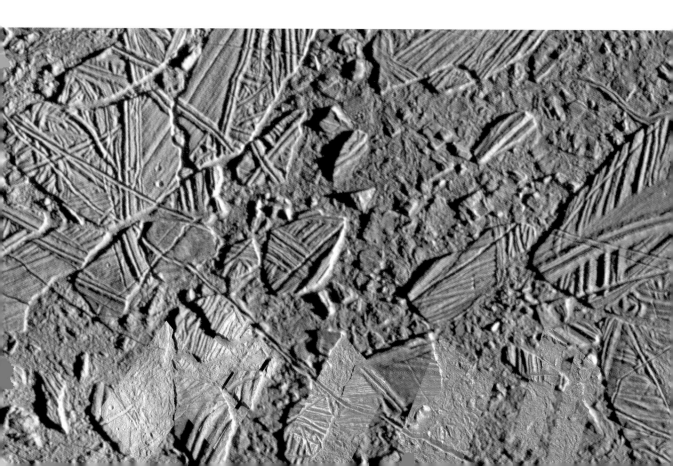

of Europa's gravity field. And Galileo's magnetometer indicated that Europa might have a magnetic field. From these data, we deduce that Europa may have a metallic core more than 1,200 kilometers in diameter, surrounded by a rocky mantle, with a layer of water at least 100 kilometers thick on top. The big question is whether that upper layer, beneath the visibly icy surface, is (or was) wholly or partly composed of liquid water.

Unfortunately, it is difficult to infer from pictures of a surface what lies below. There are suggestions of an ocean with dissolved substances below the ice everywhere on Europa. Darker, reddish-brown discolorations of the icy surface are always associated with features apparently derived from below: the blistery lenticulae, the wedge-shaped patches between separated plates of ice, the occasional smooth areas resembling frozen lakes, and the filled-in puncture holes which mark violent impact sites. But where's the proof of an ocean below? From pictures we can no more "see" the state of material ten kilometers below the ice than we can infer, without biting into it, the nature of an exotic fruit beneath its skin. (The dimensions of the lenticulae and the ability of the ice to sustain the bowl-shaped topography of the rare impact craters imply that the ice on Europa is about ten kilometers thick.) The photogeologist must interpret aerial pictures without any help from seismometers, bore hole drilling, and other geophysical tools (such as ice penetrating radar) not yet deployed on Europa.

How can we tell if Conamara's plates floated on a liquid ocean, or plowed slowly through soft ice? Glaciers, after all, flow downhill over decades and centuries. The bubbling-up from below that apparently made the lenticulae may be rising blobs of warmer ice breaking through the colder surface layer. Certainly what underlies the top few kilometers of Europa's icy crust is less strong and brittle than the cracked and ridged surface we see, but is it liquid water, brine, slush, or just warmer, viscous ice?

Many planetary geologists would bet that an actual liquid, briny ocean lies below the ice today. They envision sending orbital radar sounders, which could find the thinnest parts of Europa's crust. Later, a radioactively-warmed probe could land on the ice, and melt its way down to the warmer, perhaps liquid, layers below. Then the search for Europan life would begin. Certainly the ingredients for life are there: water, carbon-rich nutrients from infalling comets (possibly manifested by the red-brown surface stains), and a flow of energy. The tidal energy in Europa may produce volcanic eruptions and hot springs on the floor of a subsurface ocean overlying the rocky interior. The resulting chemical reactions might support life in the sunless depths of Europa, just as the Earth's internal energy sustains thriving ecosystems around sea floor volcanic vents.

But does life really exist on Europa? Our discoveries in the solar system continue to teach us a larger lesson: Nature, in its variety and complexity, can easily outwit our attempts to theorize about it. The only way to find out if another world supports life is to observe and explore it. Then we may be able to answer a prime question of astrobiology: Does life commonly originate and survive wherever an environment can support it, or is our own planet a rare exception?

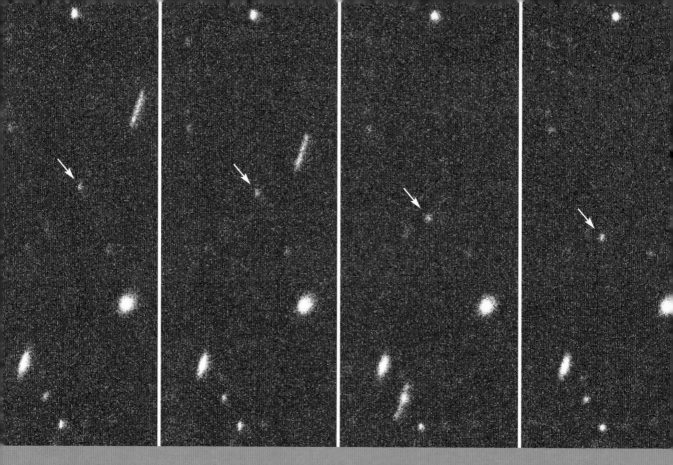

The Kuiper Belt

Jane Luu

We have long believed the solar system to be one of the
best-known and least mystifying regions of the universe. We have
understood the motions of the planets since the days of Kepler
and Newton, and the last major solar system member, Neptune,
was discovered in 1846. For a long time, the study of the
solar system has been incremental rather than revolutionary.
We have slowly built up a census of asteroids, comets, moons,
and rings. We have sent probes to the outer reaches of the
solar system, studying the planets in ever-increasing detail.

Jane Luu is an Associate Professor of Astronomy at Leiden University in Leiden, The Netherlands.

In August 1992, however, our delusions of understanding the solar system suffered a heavy blow. An entirely new class of icy objects was discovered—the Kuiper Belt. Pluto, once thought to be just a small, anomalous planet, turned out to be the largest known Kuiper Belt object (KBO), only one of a swarm of sizable bodies extending far beyond the planetary system's old boundaries. Overnight, the solar system acquired a whole new class of objects, revealing how incomplete our knowledge of it is.

The Kuiper Belt is named after Gerard Kuiper, a Dutch-American astronomer who postulated the existence of bodies beyond Pluto in 1951. However, an Irishman named Kenneth Edgeworth independently made a similar suggestion as early as 1943. The truth is that neither man possessed a physical theory to explain his hypothesis, and their speculations were little more than lucky guesses. A grand theory to explain the origin and evolution of the Kuiper Belt is still lacking today, although astronomers are slowly putting together pieces of the puzzle.

As of January 2000, more than 200 KBOs have been reported, and roughly half of them have reasonably well-determined orbits (Figure 1). These objects are estimated to be a few hundred kilometers in diameter, and almost all were discovered beyond Neptune, between 30 and 50 Astronomical Units (AU) from the Sun, (an AU is the average distance between the Earth and the Sun, or about 150 million kilometers). Even within this limited range of distances, the Kuiper Belt shows a complex structure: KBOs are not spread out uniformly in the Belt, but are clustered into subgroups, because the Belt has been extensively sculpted by the gravitational influence of the giant planets. Based on the orbits of their members, three distinct subgroups have clearly emerged: (1) the "Classical" KBOs (roughly two-thirds of the known sample), with near-circular orbits beyond 42 AU; (2) the "Resonant" KBOs (one-third of known sample), which have orbital periods that show a simple numerical ratio to the period of Neptune; and (3) the "Scattered" KBOs, which have large, highly eccentric (non-circular) and highly inclined orbits.

The near-circular orbits of the Classical KBOs have kept them relatively far away from Neptune. We therefore believe they formed more or less where they are and have survived until now. Most of the resonant KBOs are in what is called the 3:2 mean motion resonance, so named because these objects complete 2 orbits around the Sun for every 3 orbits completed by Neptune. This group includes Pluto among its members. Other objects in the 3:2 resonance have orbits very similar to that of Pluto and have acquired the label "Plutinos" to emphasize this dynamical similarity.

Four Scattered KBOs have been identified so far. The most distant of them all, 1999 CF119, has an average distance from the Sun of 115 AU and reaches 200 AU at the furthest point in its orbit. The existence of the Scattered KBO population introduces a tantalizing question: Are there undiscovered Pluto-sized bodies in the outer solar system? It would be easy for such bodies to elude most solar system surveys to date if they traveled on large and eccentric orbits like those of the Scattered KBOs. Interestingly, there is currently no observation or theoretical prediction to rule out the possibility of other "Plutos" lurking in the Kuiper Belt, still awaiting discovery.

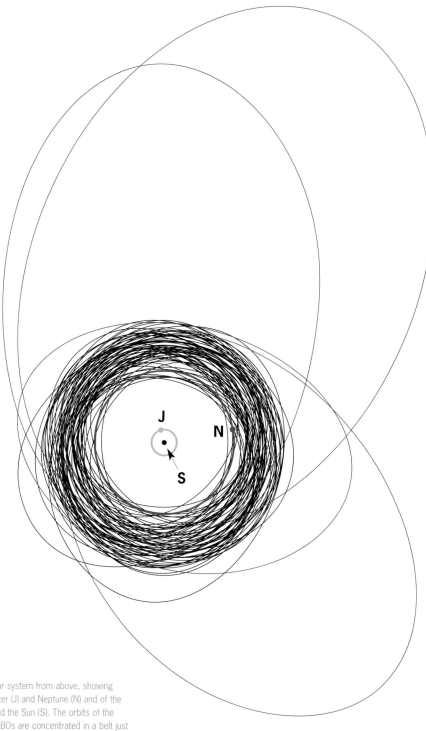

Figure 1: View of the solar system from above, showing orbits of the planets Jupiter (J) and Neptune (N) and of the Kuiper Belt objects around the Sun (S). The orbits of the Classical and Resonant KBOs are concentrated in a belt just beyond Neptune. The scattered KBOs have more eccentric orbits, four of which are shown. For scale, Neptune's orbit is 30 AU from the Sun.

KBOs travel at speeds of about one kilometer per second with respect to each other. This is much smaller than the relative velocities in the asteroid belt, but still sufficiently high that collisions in the Kuiper Belt are expected to lead to fragmentation or disruption rather than growth of objects. In other words, no growth is taking place in the present-day Kuiper Belt. In addition, collisions may crater the surface of some KBOs and excavate fresh interior material, and thus play a key role in determining the surface composition of KBOs. A well-known example of this process occurs on the surface of the Moon, where relatively recent impacts have excavated and deposited bright rays on the darker background material.

The statistics of the KBOs discovered so far imply that the 30–50 AU region currently contains roughly 100,000 bodies larger than one hundred kilometers in diameter, with a total mass of 0.1–0.2 Earth masses. In comparison, the asteroid belt contains only about 230 asteroids larger than one hundred kilometers. The Scattered population may contribute an additional 0.1 Earth masses out to about 150 AU, although this is highly uncertain. All mass estimates of the Kuiper Belt are uncertain; except for Pluto and its satellite Charon, there is no direct measurement of the size of any KBO. The sizes quoted here and elsewhere were all calculated from the known distance and the observed brightness. We have assumed that KBOs, like comet nuclei, are dark, reflecting only four percent of incoming sunlight. If KBOs are more reflective (Pluto's reflectivity is sixty percent, thanks to its surface frost), then their estimated size, and the mass of the Kuiper Belt, will have to be scaled down accordingly. However, Pluto's surface frost stems from interactions between its thin atmosphere and its surface, and it is unlikely that most known KBOs have enough gravity to sustain atmospheres.

All Kuiper Belt objects have average distances from the Sun greater than 30 AU, beyond the orbit of Neptune, showing that this planet defines the inner edge of the Belt. No KBO has been observed beyond 50 AU. Does this imply the existence of a physical edge at 50 AU? Or do only small KBOs exist beyond this boundary, which cannot be detected with current technology? The answer remains anyone's guess, and will be difficult to reach, considering the difficulty involved in finding KBOs even at the inner edge of the Belt. It is important, however, to resolve this puzzle for it bears directly on one of the most important problems in the origin of planetary systems: How far out can planetary formation occur?

We are only beginning to probe the dynamical processes that operate inside the Kuiper Belt. It appears that the Belt is a remnant from the formation of the solar system. Calculations suggest that gravitational perturbations from the giant planets must have eroded the Kuiper Belt considerably—by as much as ninety percent—over the age of the solar system, and the current population is but a shadow of its former self. How massive was the Belt initially? An accurate answer would require a detailed knowledge of the history of the Kuiper Belt, which is currently outside our grasp. It is likely that KBOs grew concurrently with the formation of Neptune, and must have finished forming at roughly the same time as this planet. Why? Because a fully-grown Neptune would have severely hindered growth in the nearby Kuiper Belt by gravitationally accelerating KBOs to the point where collisions between them are too violent for agglomeration. The formation time for Neptune is uncertain, but is estimated to be roughly 100 million years. Scientists are beginning to explore the early Kuiper Belt through computer modeling, and it appears that the original Belt must have been about 100 times more massive than its current state.

The size and mass of the Kuiper Belt establish it as a substantial and important solar system population in its own right. But scientists interested in the Belt are after much more than just adding new members to the solar system inventory. It turns out that the Kuiper Belt is intimately related to other bodies in the outer solar system and, in many cases, its discovery solves long-standing problems concerning these bodies' origins. For example, the Kuiper Belt is now accepted as the source of the short-period comets (those with orbital periods of less than 200 years). Just as the Oort cloud, a spherical cloud of comets at roughly 40,000 AU, supplies the long-period comets, the Kuiper Belt supplies a steady influx of short-period comets to the near-Earth region. Thus, by studying KBOs astronomers can in effect glimpse the solid cores of comets as they have never been seen before—in their pristine state, before they are modified by solar heating.

The Kuiper Belt also removes much of the mystery surrounding the origin of Pluto. As the most illustrious member of the Kuiper Belt, Pluto can now safely claim an origin in the Belt, to be understood in the same framework as the other KBOs. The Kuiper Belt can probably also lay claim to the Neptunian satellite Triton. The retrograde orbit of Triton (it orbits in the opposite direction to the spin of Neptune) and its physical similarities with Pluto have long hinted at a common origin with the latter. Finally, astronomers interested in exploring the very root of the solar system now have the opportunity to observe directly the remnants of the solar nebula, the original cloud of dust and gas from which the solar system formed. KBOs are by far the best-preserved fossils of the nebula, and astronomers will be busy deciphering the imprints of the solar nebula on these bodies for years to come.

Solar system astronomers are not the only ones interested in what the Kuiper Belt may reveal. Stellar astronomers quickly recognized that in the Kuiper Belt they have found a promising link between our solar system and extrasolar circumstellar disks—dust and gas disks orbiting around other stars (see the essay by Steven V. W. Beckwith in Section Two). Astronomers have long suspected that circumstellar dust disks, such as the famous one around the star Beta Pictoris, harbored planetary systems, but conclusive evidence has always been elusive. A thorough comparison of the Kuiper Belt and known circumstellar disks may provide a conclusive answer to the true nature of these dust disks. The current dust content of the Kuiper Belt is uncertain, but a rough estimate is 1,000 times less than the dust mass around Beta Pictoris. However, if the Belt was once 100 times as massive as today, perhaps its dust content was once also large enough to rival the dusty appearance of known circumstellar disks.

If the Kuiper Belt (in its current or earlier state) could be proven to be a local analog of circumstellar disks, then the presence of planetary systems inside these disks would appear likely. Thus, astronomers may one day find that, as a result of digging around their own backyard, they have stumbled upon a much bigger prize: the understanding of planetary systems as a general, Galaxy-wide phenomenon. 🪐

Gerard Kuiper and the
Trans-Neptunian Comet Belt

The census of the solar system is far from complete. When Pluto was found in 1930, it was hailed as the most distant planet. However, it turned out to be smaller than the Earth's Moon, and there remained a puzzling lack of sizeable objects beyond the orbit of Neptune. The comets, however, had long brought us hints of what lay in the farthest reaches of the solar system.

Comets are dusty snowballs on elliptical orbits around the Sun. When a comet enters the inner solar system, sunlight evaporates the surface ice and blows back a prominent tail of gas and dust. If we list comets according to their orbital periods, two distinct groups emerge. Short-period comets orbit the Sun in the same direction as the planets, and have periods less than about 200 years. In contrast, long-period comets have extremely stretched-out

elliptical orbits, with periods often more than a million years, and they enter the inner solar system randomly from all directions. Any long-period comet will be seen by us only once. Indeed the orbits of these comets must take them hundreds to thousands of times farther from the Sun than the orbit of Pluto. Where do they come from?

In 1950, the Dutch astronomer J. H. Oort proposed that all comets come from an enormous spherical cloud of objects extending almost halfway to the nearest stars and only weakly bound by the Sun's gravity. This cloud, now called the Oort cloud, is occasionally perturbed gravitationally as it moves through the Galaxy. Such disturbances hurl comets from the cloud into the inner solar system, where we can see them.

Orbit of Neptune

Oblique view of the solar system, showing the main concentration of Kuiper Belt objects beyond the orbit of Neptune. The outer edge of the Kuiper Belt is not known. The orbit of Neptune is 60 Astronomical Units in diameter.

Most astronomers immediately accepted the existence of the Oort cloud, but its origin remained unclear. Furthermore, while the theory accounted for the long-period comets, it did not adequately explain how short-period comets acquired their relatively small orbits with a preference for the plane of the solar system. It was assumed that short-period comets must be long-period comets that had passed close enough to Jupiter for its gravity to pull them into the plane of the solar system and shorten their orbits. But the idea was never verified quantitatively.

In 1951, the Dutch-American astronomer Gerard Kuiper wrote an influential paper on the origin of the solar system. Like others, he assumed that the planets had condensed from the solar nebula, a disk-shaped cloud of gas and dust in orbit around the young Sun (an idea first proposed by the philosopher Immanuel Kant in 1755). When Kuiper estimated the original amounts of material in the solar nebula at various distances from the Sun, he found that the distribution increased to a peak at the distance of the giant planets Jupiter and Saturn and then declined beyond the orbit of Neptune. His model of the solar nebula beyond Neptune had insufficient mass to form another large planet, but a smooth distribution of leftover gas and dust extended to greater distances. Kuiper calculated that this material would have condensed to form billions of small icy bodies with the size and composition of comets. He also concluded that gravitational perturbations by the planets would have thrown most of these bodies out to the distance of the Oort cloud, "where stellar perturbations would have altered their orbits once more, making them rounder and of random inclination."

But not all the comets would have been ejected to the Oort cloud. Kuiper assumed that beyond the orbit of Pluto, "remnants of the circular comet ring are still present." This hypothetical donut-shaped region of comets beyond Neptune became known as the Kuiper Belt. (It was also called the Edgeworth-Kuiper Belt, in recognition of the Irish astronomer Kenneth Edgeworth, who had earlier made a similar suggestion.)

In the mid-1980s, David Jewitt and Jane Luu began to search for the small bodies that might populate a distant Kuiper Belt. They expected that modern observing techniques would be able to render such faint objects visible. They took hundreds of pictures for many years, scouring them for any signs of faint, slowly moving objects. Finally, in late 1992, after a night of observation on the peak of Mauna Kea in Hawaii, they compared two images taken fifteen minutes apart and saw a faint object that had moved slightly through the sky. Jewitt says, "We both fell silent." After taking several more images of the same part of the sky, there could be no doubt: they had discovered the first Kuiper Belt Object (KBO). It was a small world of ice and dust, perhaps 250 kilometers across, with an orbit somewhat larger than Pluto's. To date over 200 KBOs have been found. Only the largest ones can be observed, but estimates suggest that the KBOs number in the billions.

The Kuiper Belt is the oldest surviving remnant of the original solar nebula. It represents the original source of the far more distant Oort cloud comets, as Kuiper had proposed. Furthermore, the Belt may also be the source of some of the short-period comets. Most astronomers (including Kuiper) had thought that these were long-period Oort cloud comets perturbed into short-period ones by Jupiter, but calculations have shown that this mechanism is inefficient. Instead, it was found that the gravity of Neptune can hurl enough KBOs into

the inner solar system to account for the population of short-period comets. Since KBOs already orbit in the same direction as the planets, this mechanism explains the direction of short-period comets moving around the Sun. And since most of the known KBOs already have periods of only a few hundred years (unlike the long-period comets), perturbation by Neptune can convert enough of them to account for the observed short-period comets.

The discovery of the Kuiper Belt has made profound changes in our picture of the outer solar system. Even Pluto is now regarded by many astronomers as a giant comet—the largest of the KBOs. As observations continue to bring more of these distant objects into view, we should count on new surprises.

Upsilon Andromedae

Detecting Extra-Solar Planets

David C. Black

Introduction

Nearly 500 years ago, the Polish astronomer Nicolaus
Copernicus declared that the then-prevailing view of the Earth
as the center of the universe was wrong. Instead, the
Earth was but one of several planets that revolve around the
Sun. This revolution in thought had to overcome fierce
opposition. In its futile attempt to suppress the Copernican
view, the Inquisition burned the philosopher Giordano Bruno
at the stake and subjected Galileo to house arrest.

David C. Black is the Director of the Lunar and Planetary Institute and Vice President
for the University Space Research Association Space Programs.

Today, the idea that the Earth revolves around the Sun is hardly cause for such severe reaction. Modern hypotheses as to how stars and our own planetary system form, as well as observations of newly-formed stars, suggest that planetary systems may be a rather frequent occurrence in our Galaxy. However, until we are able unambiguously to detect other planetary systems, we will not know whether the characteristics of our solar system are unique, unusual, or common. And the extension of the "Copernican Revolution"—the view that there is nothing special about our place in the universe—will remain incomplete.

The detection of other planetary systems is also important because planets are thought to be the most likely, and maybe the only, abodes of life in the universe. If so, knowledge of the existence and nature of other planetary systems is pivotal to our understanding of the possibility of life existing beyond our own planetary system.

Search Methods

Methods to search for other planetary systems can be placed in one of two categories, depending on whether planets are detected directly or indirectly. Direct methods are those that detect either reflected visible light (this is how we see the planets Venus or Jupiter from our backyards, for example), or intrinsic (e.g., thermal) radiation from the planet (this is the radiation coming from a hot or warm surface that you feel, but infrared telescopes "see"). Direct detection methods attempt to minimize or eliminate the light from a star in order to see the planets revolving around it. Indirect methods attempt to observe the "wobble" in the position

of a star due to the gravitational pull of unseen companions (possibly planets) in orbit around it. Each of these techniques tells us something different about any planets discovered.

Direct Detection

Direct detection of a planet around another star is difficult because the central star is so much brighter than the planet, and when viewed over interstellar distances, any planet will appear close to the star. A sense of the brightness of a star like the Sun, relative to a giant planet like Jupiter, a planet like Uranus, and one like the Earth, is shown in Figure 1. Each curve plots the amount of radiant energy detected from a given object, across a spectrum of wavelengths ranging from visible to infrared light. The upper

Figure 1: A comparison of the energy given off by the Sun and by three of its planets, plotted versus the wavelength of light. Each interval on the vertical axis corresponds to a factor of ten in energy. The Sun is nearly a billion times brighter than any of the planets in visible light, at wavelengths from 0.4 to 0.7 microns. Earthlike planets are brightest at wavelengths between about 7 and 20 microns, in the infrared part of the spectrum. The broad peaks in the planetary spectra at shorter wavelengths are due to reflected sunlight. The peaks at longer wavelengths are due to thermal radiation emitted by the planets themselves.

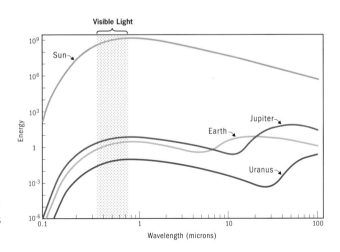

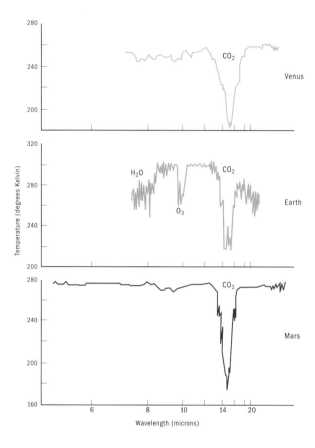

Figure 2: A comparison of the infrared spectra of Earth and two other Earthlike planets. The vertical axis represents energy received at each wavelength (labeled for technical reasons in terms of an equivalent temperature). The dips in the spectra are due to the absorption of energy at specific wavelengths by molecules in a planet's atmosphere. All three spectra show evidence of atmospheric carbon dioxide (CO_2), but only the Earth shows evidence of water vapor (H_2O) and of ozone (O_3), a surrogate for molecular oxygen.

thousands of times dimmer at infrared wavelengths. The infrared is also an interesting region of the spectrum for other reasons (Figure 2). It is here that we might observe a rich mix of spectral "signatures" from various molecules in the atmospheres of distant planets. The relative abundances of some molecules, such as oxygen (O_2), ozone (O_3), carbon dioxide (CO_2), and water (H_2O), may indicate whether a planet is habitable.

Our ability to detect other planetary systems directly is marginal with current telescopes. An interesting exception would be the detection of young Jupiter-like planets. These bodies would be hotter and therefore brighter than a mature planet, because they would still retain more of the heat acquired in their formation from compression due to gravity. This would make them easier to detect.

Indirect Detection

If a star has companions revolving around it, the star itself revolves around the center of mass, or barycenter, of the entire system. Our Sun, for example, revolves around the barycenter of the solar system, a point somewhat offset from the center of the Sun by the gravity of the planets.

Astrometric and radial velocity observations, the two most frequently used indirect detection techniques, both rely on the fact that the relatively tiny motion of a star due to the gravity from its planets can, in principle, be detected.

curve is appropriate for a body having the surface temperature of the Sun. This is an idealization, but a fairly accurate one. The portion of each planetary curve that reaches a broad peak in the visible part of the spectrum (at a wavelength of about 0.5 micron) is due to the reflected light from the central star (this is the light our eyes can see coming from Jupiter, for example). The curves for the planets also show peaks at longer infrared wavelengths, which are due to the thermal energy emitted by objects with cooler temperatures characteristic of planets.

As can be seen from Figure 1, the planets are roughly a billion times dimmer than the star at visible wavelengths, but only a few tens of

Astrometric observations detect the star's orbital motion around the barycenter in the apparent plane of the sky. In contrast, radial velocity observations detect the component of this motion along the line-of-sight to the observer. The former requires precise measurements of the position of a star over time, while the latter requires precise measurements of the spectrum of the star.

Periodic variations in a star's position or spectrum, the latter evidenced through Doppler shifts in the light from the star, reveal the presence of a companion or companions. Doppler shifts in starlight are systematic changes in the wavelengths of spectral features due to the motion of the star toward or away from the observer. When the star is going away from the telescope, the light is "redshifted" to longer wavelengths (or lower frequencies). When the star is moving toward the telescope, the light is "blueshifted" toward shorter wavelengths (higher frequencies). A similar effect occurs with sound; the pitch of a train whistle as the train approaches the observer is higher than when it rushes by or moves away.

Figure 3 indicates how the Sun would appear to "wobble" around the barycenter of the solar system if viewed from a vantage point some thirty-two light-years away. The numbers indicate the time in years. As can be seen, it takes a long time in some cases to understand the full nature of the motion (there is job security in this business!).

There is one other indirect method that has promise, namely that of photometric detection. The basic idea here is that if a companion to a star should happen to pass between an observer and the star (what astronomers refer to as a "transit"), the star will appear to dim because the companion blocks some of the light. This dimming will happen every time the companion comes around in its orbit. This

seems like an unlikely chance alignment of orbits with an observer, and it is. The likelihood that any star picked at random would be oriented just right is only a few parts in several thousand. But, if one could monitor a few thousand stars simultaneously, and if a large fraction of them had companions, then one could expect to detect a few transits per year. This approach recently succeeded in discovering a low mass companion that transits every 3.5 days across the face of its star.

Results to Date

While there have been long-term astrometric programs to search for other planetary systems, mainly at observatories such as the

Figure 3: A plot of the calculated motion of the Sun as seen by an observer looking down on the plane of the solar system from a distance of thirty-two light-years. The motion is due to the combined gravitational influence of all the planets on the Sun. The position of the Sun at various dates is indicated along the curve, starting in the year 1980. The scale of the motion is 1/1,000 of a second of arc (0.001")—equivalent to one inch seen from a distance of 3,000 miles! Observing such a small effect over many years is extremely difficult.

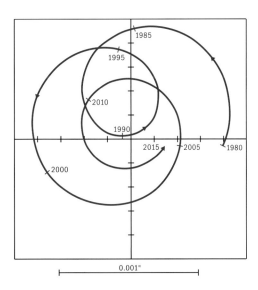

0.001"

Sproul, Allegheny, and the U.S. Naval Observatories, these have produced to date no unequivocal evidence of a planetary system in decades of searching. This negative result is due to the limited accuracy of the astrometric method rather than to the absence of planetary systems. New detectors at the Allegheny Observatory have permitted a major increase in precision, matching that of the best space-based systems, and will make it possible to detect Jupiter-like planets around nearby stars. Most of the recent activity and success in finding low mass companions to nearby stars has come from a variety of radial velocity programs in Switzerland, California, and Texas. Shown in Figure 4 is a schematic listing of twenty-seven stars around which radial velocity

studies have revealed the presence of possible planetary companions. In addition to these systems, radial velocity studies and direct searches have revealed an amazing array of objects having sub-stellar mass.

Figure 4: A schematic list of the stars that have been shown by radial velocity (Doppler) observations to have low mass companions. A lower limit for the mass of each companion is labeled in units of Jupiter's mass. The companions revolve around their stars in orbits that range from very small (at the top of the figure) to as far away from their star as the asteroid belt is from the Sun (2–3 AU). No companions have been discovered with certainty that are as far away from their star as Jupiter is from the Sun (5 AU). The star Upsilon Andromedae has three low mass companions, but further observations are needed to confirm this. Extrasolar planets are being discovered at a rate of about one per month.

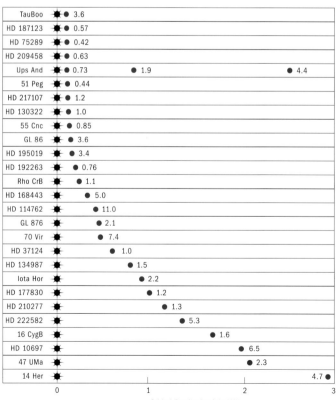

Most of these latter objects, perhaps all, are "brown dwarfs," having masses intermediate between those of stars and planets. Unlike planets, which form in circumstellar disks, brown dwarfs form in the same way as stars, by the breakup of interstellar clouds into smaller pieces due to gravitational collapse. However, brown dwarfs are not massive enough to qualify as stars. They have less than eight percent of the mass of the Sun, which is not enough for them to become hot enough to fuse hydrogen into helium and release visible light (i.e., to shine).

A key, yet unresolved, question about Figure 4, is "What are these companions?" Some of them are in elliptical orbits smaller than that of Mercury around the Sun. Many scientists have asserted that these companions must be planets, as their masses appear to be similar to, but generally somewhat more than, the mass of Jupiter, the heaviest planet in our solar system. They argue that the mechanisms of collapse due to gravitational fragmentation of interstellar clouds, thought to be responsible for the formation of stars and brown dwarfs, could not make objects this small.

However, our understanding of how these objects were formed is almost non-existent. Most of the surprise produced by their discovery is only warranted if they are indeed planets, formed by the accretion of material orbiting in a disk around a star. If the newly-discovered objects instead formed in the same manner as stars, by direct collapse from an interstellar cloud, then all their orbital characteristics are no longer surprising. The close proximity of most of these objects to their central star is consistent with the behavior of binary stars, which are systems of two stars gravitationally bound to each other. Moreover, the orbital properties such as eccentricity (the departure of an orbit from circular) of all

of the companions are consistent with the orbital properties of binary stellar systems, not planets.

Indeed, there is more than a hint that many, even most, of the companions shown in Figure 4 might be an entirely new class of object. This new type of object is more star-like in its properties than planet-like, so the question of whether we have yet detected any other planetary systems must be viewed as open. This is in contrast to statements in the popular, and even the scientific, press that extrasolar planets have been discovered.

There is at this time no unequivocal evidence that we have detected another planetary system, let alone one that is even approximately similar to our own. It is therefore premature to draw conclusions as to the frequency of occurrence of planetary systems around the multitudes of stars in our corner of the Milky Way, let alone the few hundred billion stars in our Galaxy as a whole. But it would be surprising if only the Sun had this remarkable collection of planetary companions accompanying it on its journey through space.

As discussed by Alan Dressler in Section Six, a major objective of NASA's Origins Program is the detection and characterization of other planetary systems. At least one of the future instruments that could be developed through the Origins Program would also provide the capability to assess whether planets that are discovered are habitable, or even inhabited.

We stand on the threshold of an era when humanity will complete the revolution in thought started by Copernicus some 500 years ago. Whatever the outcome of the ongoing searches, and those that will follow, our view of the universe in which we live, and our place in it, will be forever changed.

Interplanetary Impact Hazards

David Morrison

At a time when natural hazards and risks of all kinds are frequently in the headlines, scientists have identified a new hazard—one with the potential to end human civilization. We have recognized for nearly a century that the Earth orbits the Sun in a sort of cosmic shooting gallery, subject to impacts from comets and asteroids. It is only fairly recently, however, that we have come to appreciate that these impacts by asteroids and comets (often called Near Earth Objects, or NEOs) pose a significant hazard to life and property. Although

David Morrison is Director of Astrobiology and Space Research at NASA's Ames Research Center.

the annual probability of the Earth being struck by an asteroid or comet is extremely small, the consequences of such a collision are so catastrophic that it is prudent to assess the nature of the threat and prepare to deal with it.

The NEOs are usually divided into asteroids and comets. The asteroids are irregular, mostly rocky objects that come primarily from the main asteroid belt between the orbits of Mars and Jupiter. Rare collisions among asteroids can scatter some of the fragments into Earth-crossing orbits. Other NEOs, with a greater proportion of ice and other frozen volatiles in their makeup, are comets that came from the outer edges of the solar system. A few NEOs are probably old comets that have lost most of their ices, and it is hard to distinguish these from the rocky asteroids. But whatever their composition, once an NEO is on an Earth-crossing orbit, it repeatedly interacts gravitationally with the Earth or one of the other inner planets. Sometimes the slingshot effect of these encounters throws the NEO back out into the main asteroid belt, or even ejects it entirely from the solar system. Other NEOs eventually end up colliding with one of the inner planets. Of the currently known Earth-crossing asteroids and comets, probably about twenty percent will eventually strike our planet.

We are concerned about these strikes because even a small impact (from an astronomical perspective) can precipitate an environmental disaster of global proportions. Billions of tons of hot rock dust and vapor are ejected; suspended in the atmosphere, the dust can block sunlight from reaching the surface, and in the weeks after an impact this dust pall will spread over the entire planet.

The consequences of an impact depend strongly on the size and energy of the NEO. Fortunately, the Earth's atmosphere protects us from most NEOs smaller than a modest office building (50-meter diameter), impacting with the explosive energy of five megatons of TNT. From this size up to about 1-kilometer diameter, however, an impacting NEO can do tremendous damage on a local scale. Such damage includes large tsunamis from oceanic impacts that might devastate thousands of kilometers of coastline. Impacting bodies larger than about two kilometers would deliver the explosive energy of a million megatons, which would produce severe environmental damage on a global scale. The probable consequence would be an "impact winter" with loss of crops worldwide and subsequent starvation and disease. Even larger impacts can cause mass extinctions, like the one that ended the Age of the Dinosaurs 65 million years ago.

We have some experience with impacts of smaller size. United States defense surveillance satellites frequently detect the atmospheric entry of objects up to tens of meters in diameter. The largest reported such event, on February 1, 1994, penetrated to within twenty-one kilometers of the surface before exploding in the atmosphere with the energy of approximately one hundred kilotons, about eight times that of the Hiroshima nuclear bomb. About one hundred times larger was the explosion of the asteroidal fragment that entered the atmosphere over the Tunguska River region of Siberia on June 30, 1908. The resulting airburst at an altitude of eight kilometers has been estimated at about fifteen megatons of explosive energy, similar in magnitude to a large "city-killer" nuclear bomb. More than 1,000-square kilometers of forest was devastated, and the atmospheric pressure wave was measured on barometers around the world. Finally, the impact that ended the Cretaceous era 65 million years ago had an energy estimated at roughly 100 million megatons. As is now widely known, the resulting environmental catastrophe led to the extinction of more than half the species on Earth, including the dinosaurs.

The risk from cosmic impacts increases with the mass of the projectile. Objects as small as the meteorites we see in museums pose only a negligible threat, with no authenticated fatality due to being struck by such a space rock. Even the large Tunguska explosion actually killed only one person. It turns out that the greatest risk is associated with objects large enough to perturb the Earth's climate, leading to massive loss of food crops and possible breakdown of society. Such global catastrophes are qualitatively different from other more common hazards that we face (excepting nuclear war), because of their potential effect on the entire planet and its population. Cosmic impacts represent the extreme example of a natural hazard of very low probability but extremely great consequences.

In order to deal with this potential threat, we must discover whether there are actually any NEOs on a collision course with Earth. Until we have surveyed the NEO population, we cannot rule out the possibility the Earth might suffer a major impact in the next century, the next decade, or even next year.

At the request of Congress, NASA has sponsored two studies of ways to discover Earth-crossing asteroids or comets before they pose any direct threat to the planet. The first study, which I chaired, was completed in 1992 and developed the idea of an international Spaceguard Survey of specially-designed ground-based telescopes to detect and catalogue all asteroids larger than one kilometer in diameter within the next twenty-five years. The follow-up study, chaired by the late Eugene Shoemaker of Lowell Observatory and completed in 1995, concluded that advances in astronomical

imaging systems could allow the Spaceguard Survey to be completed in just ten years at a total cost of less than 100 million dollars. The proposed Spaceguard Survey would identify any large objects that could hit the Earth through a systematic search that effectively monitors a large volume of space around our planet and detects NEOs as their orbits repeatedly carry them through this volume of space.

Although the full Spaceguard plan has not yet been implemented, several teams consisting of a few dozen astronomers worldwide are surveying the sky with electronic cameras to find NEOs. The most productive NEO surveys in 1999 were: the LINEAR search program of the MIT Lincoln Lab, carried out in New Mexico with NASA and U.S. Air Force support; the NEAT search program in Hawaii, carried out jointly by the NASA Jet Propulsion Lab and the U.S. Air Force; the Spacewatch survey at the University of Arizona, funded by NASA and a variety of private grants; and the NASA-supported Lowell Observatory NEO Survey. Other searches in the U.S., France, Japan, and China also contribute to discovery of NEOs, while additional astronomers (many of them amateurs) follow up the discoveries with supporting observations in order to determine accurate orbits. We have already cataloged more than 800 NEOs, including nearly 400 of one kilometer or greater diameter.

None of the known NEOs is on a collision course with Earth. All known NEOs and their predicted future positions are openly available to everyone with access to the Internet (see http://impact.arc.nasa.gov and http://neo.jpl.nasa.gov). The problem is that astronomers have discovered only about twenty percent of even the larger NEOs, and we have no way of predicting the next impact from an unknown object. But we can calculate the odds: on average, a one-million megaton (globally catastrophic) NEO collides

Terrestrial impact frequency

The diagonal line plots the expected average time interval between asteroid or comet impacts on Earth versus the energy of the explosion. Because large interplanetary objects are rarer than small ones, the average time interval between impacts increases with the energy of the event. The explosive energy is expressed in megatons of TNT. The 1908 Tunguska explosion over Siberia delivered an energy of about fifteen megatons. An impact of that energy is expected on Earth every few centuries. The Cretaceous/Tertiary impact (K/T event) that wiped out the dinosaurs had an energy of about 100 million megatons. An impact of such magnitude is expected about once every hundred million years. The energies of the Hiroshima bomb (thirteen kilotons) and the largest hydrogen bomb (sixty megatons) are shown for comparison.

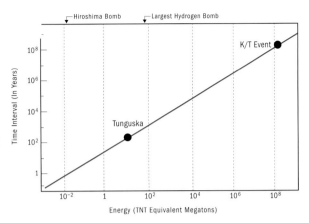

with the Earth once or twice per million years, producing an "impact winter" that would kill a substantial (but unknown) fraction of the Earth's human population. Reduced to personal terms, this means that you have about one chance in two million each year (or a lifetime chance of about one in 20,000) of dying as a result of such a catastrophe. Such statistics are interesting, but they don't tell you, of course, when the next catastrophic impact will take place—next year or a million years from now. That is the rationale for the Spaceguard Survey.

The search for NEOs and predictions of their future paths are of more than academic interest, because we are developing the technology to protect our planet. An NEO impact is the only major natural hazard that we might potentially protect ourselves against by deflecting or destroying the threat before it strikes. The first step in any program of planetary defense is to find the NEOs; we can't protect against something that we don't know exists. We also need a long warning time, at least a decade, to send spacecraft to intercept the object and deflect it.

Many defensive schemes have been studied in a preliminary way, but none in detail. Presumably, our present level of space technology would permit us to carry out multiple spacecraft rendezvous with the threatening NEO, examine it carefully, and use one or more nuclear

explosives to nudge it into a new orbit that would miss the Earth. Even if there were insufficient warning time for such an active defense, warning of the time and place of an impact would at least allow us to store food and supplies and to evacuate regions near ground zero where damage would be the greatest. In addition, we wish to learn more about the chemical and physical nature of the NEOs. To explore the asteroids, the NASA Galileo spacecraft flew past the main-belt asteroids Gaspra and Ida en route to its 1995 rendezvous with Jupiter, and in 1996 we launched a dedicated asteroid mission called NEAR (Near Earth Asteroid Rendezvous) which went into orbit around the asteroid Eros in the spring of 2000 for an extensive year-long investigation. Imaging of near-Earth asteroids is also possible, using powerful planetary radars supported by NASA at Goldstone, California, and Arecibo, Puerto Rico.

Today, the impact risk is widely recognized, at least at the level of science-fiction novels and Hollywood films. In addition, the discovery rate of NEOs has increased dramatically, by a factor of 5 between 1997 and 1999, primarily as the result of a search partnership between NASA and the U.S.A.F. Space Command. With continued public interest and government support, we may soon have the capability to protect the Earth against this potential cosmic threat to our future.

Contemporary broadsheet depicting the fall of a meteorite at Ensisheim in Alsace on November 7, 1492, with verses by Sebastian Brant. Chladni cited this event.

Ernst Chladni and Rocks from the Sky: Profile

Ernst Chladni (1756–1827), German physicist and founder of the science of meteoritics

When a rocky or iron mass from the asteroid belt plunges into the Earth's atmosphere, it can produce a fireball streaking across the sky and sonic booms. If the mass or its fragments survive, they fall to the surface. Many people may see the fireball and hear the explosion, but only rarely are people close enough to see a meteorite hit the ground.

For thousands of years, people in all parts of the world had reported such phenomena, which they often regarded as divine portents. The scientifically-inclined suggested that the rocks were lofted by distant volcanoes or hurricanes, or perhaps congealed from atmospheric vapors under the influence of lightning. No one imagined that their origin was outside the Earth. By the late eighteenth century, scientific opinion, reflecting the spirit of the Enlightenment, questioned the credibility of wonders reported by untrained observers. The whole subject of rocks from the sky was dismissed as common superstition.

Then in 1794, the German physicist Ernst Chladni published a small book asserting that masses of iron and of rock actually do fall from the sky, producing fireballs when heated by friction with the air. He concluded that they must be cosmic objects, perhaps debris ejected from planets by explosions or collisions. Reaction to the book ranged from skepticism to ridicule. How could there be rocks in space? Aside from the stars, planets, moons, comets,

and perhaps some vapors arising from their atmospheres, everyone knew that space itself was empty. Aristotle and Newton had said so. And yet Chladni was right. Today he is regarded as the founder of meteoritics.

Ernst Florens Friedrich Chladni (1756–1827) was fascinated by science and music from a young age, but was trained as a lawyer at the insistence of his father, a professor of law in Wittenberg, Germany. Later, Chladni was able to pursue scientific research. He discovered a way to make sound waves visible, by sprinkling powder on a plate of metal or glass and rubbing the edge of the plate with a violin bow. The vibrations of the plate made the powder accumulate in symmetrical patterns, now called Chladni figures. These figures revealed the modes of sound wave vibrations in the solid body.

Chladni's interest in meteorites was stimulated in 1793 by a conversation with Georg Lichtenberg, professor of physics at the University of Göttingen. Lichtenberg had witnessed a fireball and thought that such phenomena might be due to cosmic bodies entering the Earth's atmosphere. Chladni began his investigation by searching the literature for eyewitness accounts of fireballs and rocks from the sky. During three weeks in the university library, he compiled what he felt were the most reliable eyewitness reports. The list included twenty-four fireballs and eighteen fallen rocks reported from various countries over many centuries. The similarities of these accounts impressed Chladni, whose legal training had prepared him to evaluate eyewitness testimony. He concluded that the witnesses must have been describing a real physical phenomenon.

Chladni found numerous cases in which fireballs were followed by the fall of rock or iron masses to the ground. For example, a lump of iron weighing seventy-one pounds fell from the sky over Croatia in 1751. It was sent to the Imperial Natural History Cabinet in Vienna, together with sworn testimony by seven witnesses in widely separated towns who described a spectacular fireball in the sky and loud explosions.

By analyzing the descriptions of fireballs, Chladni was able to estimate the speeds of the rocks entering the atmosphere. These speeds were enormous, faster than could be produced by the Earth's gravity alone, and were only possible for objects of cosmic origin. Another piece of evidence was the scorched appearance of the rocks themselves. They had been heated enough to melt their outer layers.

When Chladni published his book in 1794, many scientists immediately dismissed the work because it relied on eyewitness accounts. However, events in the next few years swung the weight of opinion in Chladni's favor.

Two months after the book was published, a large cloud of smoke appeared in the sky near Siena, Italy. The cloud, sparking and booming, turned bright red and stones fell to the ground. Some of the stones were recovered and descriptions of the event were published and widely discussed.

A year later a 56-pound rock fell near Wold Cottage, England. Witnesses reported the sound of an explosion from the air. One farmer actually saw the black rock hit the ground only thirty feet away, dousing him in mud.

These and similar events convinced Sir Joseph Banks of the Royal Society, an organization of leading scientists, that an investigation was warranted. He asked Edward Howard, a young chemist, to analyze the chemical composition of the alleged rocks from the sky. Howard read Chladni's book and other accounts and began acquiring samples of the stones and iron masses. Working with the French mineralogist Jacques-Louis de Bournon, he made the first thorough scientific analysis of meteorites. The two scientists found that the stones had a dark shiny crust and contained tiny "globules" (now called chondrules) unlike anything seen in terrestrial rocks. All the iron masses contained several percent nickel, as did the grains of iron in the fallen stones. Nothing like this had ever been found in iron from the Earth. Here was compelling evidence that the irons and rocks were of extraterrestrial origin. Howard published these results in 1802.

Meanwhile, the first asteroid, Ceres, was discovered in 1801, and many more followed. The existence of these enormous rocks in the solar system suggested a plausible source for the meteorites. Space wasn't empty after all.

Finally, in 1803, villagers in Normandy witnessed a fireball followed by thunderous reverberations and a spectacular shower of several thousand stones. The French government sent the young physicist Jean-Baptist Biot to investigate. Based on extensive interviews with witnesses, Biot established the trajectory of the fireball. He also mapped the area where the stones had landed: it was an ellipse measuring 10 by 4 kilometers, with the long axis parallel to the fireball's trajectory. Biot's report persuaded most scientists that rocks from the sky were both real and extraterrestrial. The science of meteoritics, the first-hand study of samples from other worlds, was finally launched.

Section Two: Stars

The Crab Nebula is the debris from a stellar explosion seen on July 4, 1054. At the center lurks a tiny neutron star spinning thirty times per second. European Southern Observatory.

Introduction Steven Soter

What is a star? Think of it as an enormous nuclear fusion reactor confined by gravity. A typical star is gaseous throughout and composed almost entirely of hydrogen and helium, the two lightest elements in the universe. A star's gravity compresses its interior, heating the central core to millions of degrees. At such temperatures, hydrogen nuclei (protons) can fuse together to make helium nuclei. This nuclear reaction liberates energy in the form of high-energy light (X-rays). The energy slowly diffuses outwards through the star's overlying layers. By the time it reaches the relatively cool surface (at a few thousand degrees), much of the light has been transformed into visible wavelengths. And so the stars shine.

The Pleiades is a cluster of several hundred stars about 400 light-years away. Wispy clouds of gas and dust reflect light from the brightest stars.

Our Sun is a rather typical star of medium size, about a hundred times the diameter of the Earth. It dominates our daytime sky only because we are so close to it, about a hundred solar diameters away. The other stars are other suns. They appear so faint only because they are millions of times farther away than the Sun. Many of them also have planets. About two-thirds of all stars are actually members of double or multiple star systems, bound together by their gravity. Our Milky Way Galaxy contains several hundred billion stars, but only a few thousand of them are close enough and sufficiently luminous for us to see with the naked eye.

Astronomers observe various types of stars, which span a wide range of mass, size, temperature (which determines color), power output, and variability. Many of these characteristics change as a star evolves through its life cycle—birth, maturation, decline, and death—but stars last far longer than a human lifetime or even human history. So astronomers cannot keep track of the life cycle of a given star. Rather, they observe a vast population of stars at the same brief instant of cosmic time and try to figure out how its members are related by species and age. It is rather like looking at an enormous but static photograph of wild animals on a savanna. By careful study, we may be able to sort out how many species are present, which individuals belong to which species, and how animals of the same species change their size and appearance with age. Much of modern stellar astronomy involves interpreting the big "snapshot" to understand the relationships and life cycles of the stars.

The life span and average luminosity, or power output, of a star depend mainly on its mass. The lower the mass, the more slowly the star consumes its fuel and the longer it lives.

Massive stars have more fuel, but they burn through it at higher temperatures and thus much more rapidly than low mass stars. A star like the Sun has a life span of about 10 billion years. A star half as massive will live about twenty times longer, while a star three times as massive will live only one-twentieth as long as the Sun.

A typical star spends most of its life quietly fusing hydrogen, at a rate that maintains a nearly constant surface temperature and luminosity. During this time, the star is in a stable balance between the hot gas pressure that pushes its matter outward and gravity that pulls it inward. But when the nuclear fuel runs out, the star begins to collapse under its own weight.

What happens next depends on the mass of the star. When a low mass star like the Sun exhausts its hydrogen fuel, the helium-rich core contracts until it becomes hot enough to fuse the hydrogen in the region surrounding the core and then to fuse the helium in the core itself, to make carbon and oxygen. By tapping these new sources of nuclear energy, the star heats up again and puffs up its outer layers. The result is a hugely bloated red giant star, which slowly sheds much of its substance into space. The core of the red giant is now a dense white dwarf star, about the size of the Earth. Further instabilities in the helium-burning zone around the core finally blow off the outer layers of the red giant, forming expanding shells of gas. The hot white dwarf illuminates the shells with ultraviolet light, making them fluoresce in brilliant colors. The result is a so-called "planetary nebula," among the most beautiful objects in the universe.

In stars much more massive than the Sun, nuclear fusion can produce elements even heavier than carbon and oxygen, such as silicon and iron. When they exhaust their nuclear fuel, such massive stars explode catastrophically as supernovas. They blow most of their substance into space, leaving behind a tiny superdense neutron star or an even denser black hole. An even more powerful kind of supernova is produced when a white dwarf star acquires enough mass from an orbiting companion star to become violently unstable. In this case, according to theoretical calculations, the white dwarf becomes a giant nuclear bomb, blowing itself completely apart and leaving no remnant.

All cases of stellar death involve the shedding of matter into space. The process enriches the interstellar medium with the heavy elements produced by nuclear fusion in the stars. Giant clouds of interstellar gas and dust are the recycling centers for the stars in the Milky Way and other galaxies. Gravity causes the densest parts of such clouds to fragment and collapse, giving birth to batches of new stars. The spectacular Orion Nebula is one of the nearest giant interstellar clouds in our Galaxy. Within it, we can observe stars in the act of formation.

As a concentration of gas and dust collapses to form a star, some of the material has sufficient lateral motion that centrifugal force prevents it from falling into the star. Instead it collapses into a flat rotating disk around the young star. Astronomers now have good observational evidence that many, if not most, young low mass stars are surrounded by such circumstellar disks. Theorists have long assumed that our planetary system must have formed from such a disk around the young Sun. If circumstellar disks are as common as suggested by the new observations, then solar systems might be quite common as well.

For many years, astronomers assumed that collisions between stars are too rare to have any observational consequences. Ordinary stars, which glide past one another in their orbits around the Galaxy, are so far apart that they will almost never collide. Only recently did anyone appreciate that conditions differ greatly within dense globular clusters of stars. Our Milky Way Galaxy has some 150 such clusters distributed in a spherical "halo" around its center. The concentration of stars within a globular cluster is a few thousand times higher than in the Galaxy as a whole, so such stars are a few thousand times more likely to suffer collisions.

Various collisions involving the same or different kinds of stars are possible, and this gives rise to a wide range of phenomena. For example, if two old sunlike stars collide, they can fuse together to form a single star with twice the mass. The new star, with more available hydrogen fuel, burns hotter and brighter than the old and nearly exhausted stars typical of the cluster. Such "born again" stars have actually been observed. Other surprises are no doubt in store in the vigorous field of stellar astrophysics.

To explore the lives of the stars, we pose the following questions:

How and where are stars born?

C. Robert O'Dell, Distinguished Research Professor of Physics and Astronomy at Vanderbuilt University, examines the Orion Nebula, a spectacular giant interstellar cloud, where he has discovered stars and perhaps solar systems in the process of being born.

What can circumstellar disks tell us about the common origin of stars and planets?

Steven V. W. Beckwith, Director of the Space Telescope Science Institute and a Professor of Physics and Astronomy at The Johns Hopkins University in Baltimore, describes the disks of dust and gas in orbit around many young stars. These appear to be solar systems in formation.

How and why do stars die? What are planetary nebulae?

Adam Frank, Assistant Professor of Astrophysics at the University of Rochester, examines the planetary nebulae and the spectacular fates of quite ordinary stars.

Why are collisions between stars more common than once thought? What happens when stars collide?

Michael M. Shara, Curator and Chair of the Department of Astrophysics at the American Museum of Natural History in New York City, discusses the surprising news that collisions between stars are actually rather frequent. The results of such encounters are more surprising still.

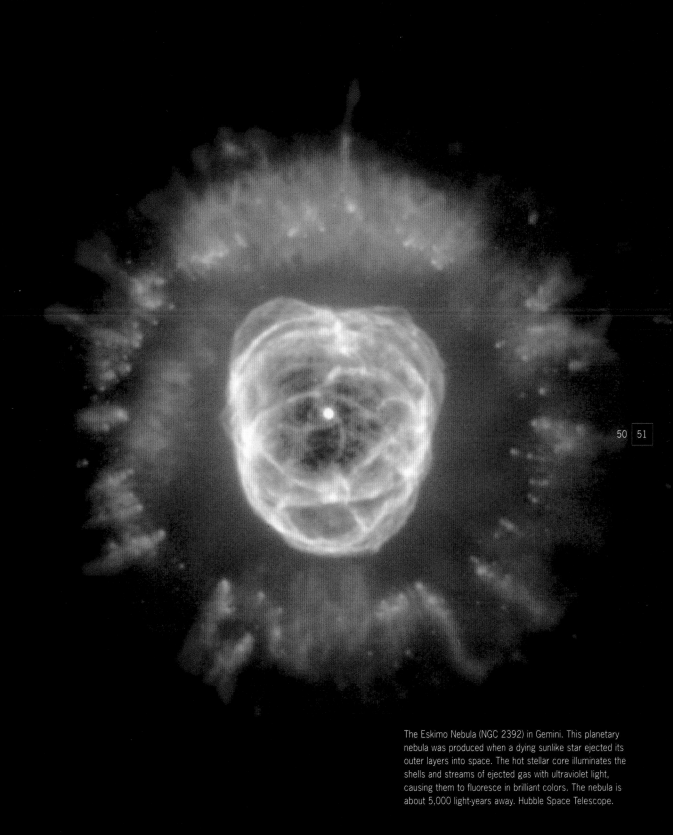

The Eskimo Nebula (NGC 2392) in Gemini. This planetary
nebula was produced when a dying sunlike star ejected its
outer layers into space. The hot stellar core illuminates the
shells and streams of ejected gas with ultraviolet light,
causing them to fluoresce in brilliant colors. The nebula is
about 5,000 light-years away. Hubble Space Telescope.

The Orion Nebula: A Stellar Nursery

C. Robert O'Dell

An examination of our neighborhood in the Milky Way Galaxy is much like an afternoon walk through a forest. There we see trees and other plants in various stages of their life cycles. We know there are factors that favor the growth of one type of tree and not another, and that the large specimens of a species are related to the small ones, but the changes that occur in the forest operate on a much longer time scale than the stroll. It would be hard to understand the life cycles in the forest based only on what we could observe in such a brief time.

C. Robert O'Dell is Distinguished Research Professor of Physics and Astronomy at Vanderbilt University.

Figure 1: The Great Nebula in Orion. This mosaic of fifteen images from the Hubble Space Telescope covers the inner 2.5 light-years of the nebula. The cluster of massive bright stars at the center (the Trapezium) illuminates the surrounding gaseous surface of the nebula.

We have the same challenge when we examine the myriad types of stars in our Galaxy. During the middle half of the twentieth century, astronomers came to realize that all stars go through similar life cycles, starting with formation, followed by a long stable period of maturity, and a final episode of enhanced brilliance before collapse. The primary factor determining what a star will be like during its lifetime is the amount of mass it has at birth. After that, its appearance and energy output will depend only on how far along the star is in its life cycle when we observe it.

In spite of our good basic understanding of these objects once they became the gravitationally bound, self-luminous bodies we call stars, the process of how stars actually form is still the subject of current research. Until only recently, our ability to describe where the stars come from was little better than speculation. However, with the advent of new observing techniques, which span the wide range from X-rays, through optical and infrared light, and right out to the end of the radio spectrum, we have begun to remove much of the mystery from the subject of star formation. Moreover, we are lucky to find that, like an acorn sprouting into a seedling, the processes occur so rapidly that time-lapse imaging can show them in action.

Often Nature seems to be teasing us, sometimes by giving tantalizing but unrevealing glimpses of star formation, at other times shrouding her secrets in veils of darkness. But she has also granted a most wondrous boon. This gift is the Orion Nebula (Figure 1), the bright core of a center of star formation lying only 1,500 light-years away and perched high above most of the obscuring dust lying in the plane of our Milky Way Galaxy. The Orion Nebula invites close examination, as it has since being first recorded by Nicholas Peiresc in 1610. This nebula is associated with the middle star in the sword of the beautiful winter constellation of Orion, the Hunter. Even a small pair of binoculars reveals that this one "star," called Theta Orionis, is actually a grouping of some seven bright stars.

But modern telescopes reveal that Theta Orionis is actually a cluster of some 700 stars. The seven most luminous stars are but the bright blue tip of a distribution of stellar brightnesses. Most of the stars have low mass. These low mass stars are quite cool and are therefore quite red, dotting the field with little rubies. Four of the seven bright stars are quite close together, in the general shape of a geometric figure called a trapezoid. This group is called the Trapezium and the associated cluster of stars is often called the Trapezium Cluster. The brightest stars of the Trapezium provide most of the illumination for the Orion Nebula.

If we measure the apparent brightness and the distance of any star, we can calculate its luminosity, or power output. But the luminosity of any incandescent body is determined by its size and temperature. Astronomers have long known how to calculate the surface temperature of a star by analyzing the spectrum of its light. Therefore, once we know the temperature and the luminosity of any star, we can determine its size.

The study of stars in the solar neighborhood indicates that most stars are within a factor of five of the size of our Sun. They spend almost all of their lifetime at a constant luminosity and surface temperature. If you plot the data for the nearby stars on a graph with temperature as the horizontal axis and luminosity as the vertical axis, you will find that most stars fall along a narrow diagonal band called the Main Sequence (Figure 2). All the stars on the main sequence are in a lengthy midlife phase of stable nuclear energy production. Such stars "burn" their

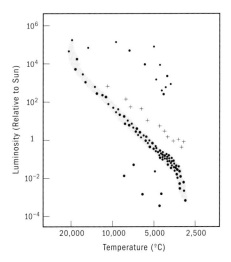

Figure 2: A schematic plot of luminosity versus surface temperature for stars in the solar neighborhood (black dots) and for stars in the Orion Nebula (plus signs). The "main sequence" of mature hydrogen-burning stars is shaded. The very young stars of the Orion Nebula lie above and are moving down toward the stable main sequence. For historical reasons, temperature (in degrees Celsius) in such a graph increases from right to left.

hydrogen fuel as their gravity squeezes them tightly enough to release nuclear energy from their cores. They are in a state of thermostatic equilibrium that can be sustained as long as sufficient fuel remains.

The relatively few stars that are plotted in other parts of the diagram are either in the formative or final stages of life. Since those stages are brief compared to the stable main sequence phase, those parts of the diagram are sparsely populated.

We don't see the standard main sequence relationship when we plot the stars of the Trapezium Cluster on such a diagram. Most of them fall above the main sequence because they are more luminous than main sequence stars at the same temperatures. This means that the Trapezium stars must be significantly larger than their main sequence counterparts. We now know the reason: these stars have been caught in the act of contracting to their main sequence destinations. They are larger than stable main sequence stars because they are younger.

The excess luminosity of a young star diminishes with time as it contracts and approaches the main sequence. The theory of star formation tells us how long this should take. Astronomers can use the observed

distribution of luminosity and surface temperature in the stars of a young cluster as a kind of clock. When they read it, they find that the Trapezium Cluster is only about one-half million years old, a tiny fraction of the ten billion year age of our Galaxy. The mere existence of such young stars means that star formation is a continuing process of regeneration.

The nurseries for these very young stars are gigantic clouds of interstellar gas, which are mostly reprocessed material ejected from dying older stars. Possessing total masses of almost a million Suns, these clouds are held together by their own gravity and become so dense that optical light cannot penetrate their interior. Molecules form from the individual atoms, probably by an exotic chemistry occurring on the surface of the dust grains mixed in with the gas. Once the density of the molecular gas becomes sufficiently high, the giant molecular clouds begin to fragment into smaller pieces, each of which is strongly bound together by its own gravity and begins collapsing to become a star.

The stars in the Trapezium Cluster have literally been caught in the act of formation. There we see a few of them have reached the stable main sequence, but most are still contracting. Moreover, we see the leftover material. This material is rendered visible as the beautiful Orion Nebula, because the leftover gas is made to glow by the hottest star in the Trapezium. The mechanism for producing the nebular light is a form of fluorescence, much like illuminating a white shirt with a barely visible ultraviolet lamp. In this case, the ultraviolet source is that Trapezium star and the resulting light comes from the gas that didn't make it into one of the collapsing stars. This mechanism of fluorescence is so efficient that it is hard to see many of the faint stars because they are lost in the glare of the bright nebula.

Based on observations, we have actually built a three-dimensional model of the cluster and the nebula and found that the cluster is in the center of a cavity in the parent molecular cloud. This suggests that the building block material has been used up there and that most of the residual material has been blown away, leaving the young stars fully visible to us.

Close examination of those young stars with the Hubble Space Telescope has revealed that many of them are surrounded by rotating disks of gas and dust. This discovery resolves a fundamental question about how stars form. We know that the pre-collapse cloudlets must be slowly rotating. As gravitational collapse progresses, a forming star must spin faster and faster in the same way that a rotating diver spins more rapidly by tucking up into a tight ball. The theoretical problem has been that if this spinup with decreasing size goes too far, then eventually the outward centrifugal force of rotation would equal the inward gravitational force of contraction, and the process of star formation would stop before an actual star was created. The mass would never become dense enough to heat up and burn hydrogen.

What we see in the Trapezium Cluster is that the forming stars cleverly circumvent this centrifugal barrier by leaving the most rapidly spinning material behind in a thin rotating disk. The central component of the gas can then go on collapsing to become a star. The spinning material left behind can become the building blocks for planets. Exactly this same kind of process must have occurred when our Sun formed some 5 billion years ago. This is why the planets in our solar system today share a common orbital plane around the Sun, called

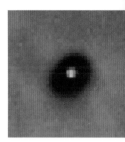

Figure 3: (top) A teardrop-shaped cloud of gas and dust surrounding a young star in the Trapezium Cluster.

Figure 4: (bottom) A young star shining through a surrounding cloud of gas and dust, seen in silhouette against the nebula. The star looks red through the dust, an effect similar to a sunset.

the ecliptic. The same processes that make the Orion Nebula so visible also illuminate the protoplanetary disks around the young stars in the Trapezium Cluster, some brightly illuminated by fluorescence and others seen in silhouette against the glow of the nebula. Figure 3 shows a disk viewed almost edge on, so that the central star is obscured, but its outer gas is illuminated by the hottest of the Theta Orionis stars. Figure 4 shows the first of the silhouette objects, where we see the forming star surrounded by its protoplanetary disk. We cannot say how many of these disks will succeed in forming planets, but we can say that almost every star has the ingredients on hand.

More recent Hubble Telescope images have shown that young stars not only shed the most rapidly spinning material in disks but also spew out beams of gas in opposite directions along the star's axis of rotation. These polar jets have "hypersonic" velocities and remain almost invisible until they run into nebular material. At that point they create energetic and luminous shock waves that then dance out ahead of the jet. The Hubble Telescope has sufficient resolution that we can see the gas streaming out in the jets, and we have been observing sufficiently long that we can see the motion of the shock waves. Rather than an ageless picture of the sky, we see a continuously changing and evolving region.

We have been lucky that nature has caused all this to occur on the side of the giant molecular cloud that faces us, so that we can see what is going on. It is like seeing our solar system as it was at its origin. Who would have imagined it?

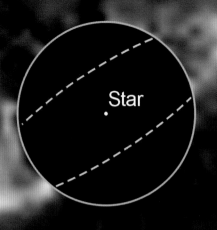

Circumstellar Disks and the Prevalence of Life

Steven V. W. Beckwith

The planets in our solar system are neatly aligned in their orbits, each path lying almost in the same plane. Each planet pirouettes on its axis and, with the exception of Uranus, the equators are roughly aligned with the common orbit plane. The plane of Saturn's rings and the orbits of many of the moons of the planets fall in the same pattern.

Steven V. W. Beckwith is the Director of The Space Telescope Science Institute and a Professor of Physics and Astronomy at The Johns Hopkins University in Baltimore.

The rough alignment of the planetary orbits was known to the ancients, and by the eighteenth century had to be part of any workable cosmogony, as theories of the origin of the solar system were then called. The Prussian philosopher Immanuel Kant and the French mathematician Pierre-Simon Laplace, inspired by Charles Messier's discovery of interstellar nebulas (clouds), independently proposed that the planets were born in a huge, swirling cloud of gas and dust. The cloud collapsed under its own gravity with most of it falling into the center to create the Sun. But parts of the cloud had enough rotation to keep them from falling into the Sun, and those parts collapsed onto a common rotating plane. Kant and Laplace envisioned the plane as an enormous circular disk, with the gas close to the Sun moving rapidly in orbit and the gas far away moving quite slowly, according to the laws of orbital motion. The planets were born in that rotating disk, which accounts for why they share nearly the same orbital plane and rotational direction. Today, such a structure is called a circumstellar disk.

Our best modern theory of solar system birth is almost identical in its main characteristics to the ideas of Kant and Laplace. As such, it is probably the oldest theory in astrophysics that has remained viable without substantial revision. We envision the early solar system, about 4.5 billion years ago, as a circumstellar disk around the young Sun. The disk contained mostly hydrogen, about seventy-five percent by mass, plus twenty-four percent or so of helium. The rest, a trace that included all the heavier elements, resided in dust and ice grains. Although dust and ice made up only about one percent of the primitive solar disk, they were the essential ingredients necessary to form the planets and provide the chemical building blocks for life.

As the gas and dust grains orbited the Sun, random motions caused the dust grains to collide with one another. The slower collisions let the grains stick together, growing from sub-micron-sized particles into pebbles and eventually large rocks and snowballs. This growth probably occurred rather quickly, so that asteroid-sized bodies were assembled in only a hundred thousand years or less. By the time these bodies, called "planetesimals," were ten kilometers across, they were able to attract more solid material through their own gravity, which accelerated the assembly process. Mergers between planetesimals eventually built the inner terrestrial planets Mercury, Venus, Earth, and Mars. In the outer solar system, the solid planetesimals became the seeds massive enough to attract the abundant hydrogen and helium gas, and the gas giants, Jupiter, Saturn, Uranus, and Neptune, were born.

Interstellar dust contains carbon, oxygen, silicon, and iron, among other elements, all of which are manufactured in stars and nowhere else. The biblical phrase "ashes to ashes, dust to dust" is given literal meaning by modern science, for dust is the ash of stellar fires, and we are built of atoms from interstellar dust. The assembly of this dust into a coherent structure was the remarkable event that made possible the creation of the solar system and the emergence of life on Earth.

After the 10 to 100 million years it took to create the planets we see today, a swarm of small rocks and asteroids remained in the disk. The young planets endured another few hundred million years of collisions with these bodies. Thus, the early Earth was a very inhospitable place. Asteroid impacts of the sort that much later may have killed the dinosaurs were then quite common. Many of the impacts had enough energy to vaporize the outer few meters of the Earth's surface. These events boiled away oceans and produced an atmosphere of vaporized rock at several thousand degrees that remained hot for thousands of years before condensing out on the surface. Any life that arose during this time was either protected from these harsh conditions deep underground or was quickly extinguished following an impact.

If our solar system were unique, the concept of a circumstellar disk would have remained merely a curious theoretical requirement needed to explain its origin. But in fact, disk-shaped structures are not unusual in the Galaxy. Many interstellar clouds arise as collections of matter blown out from old stars. These clouds gather together until the pull of their own gravity causes them to collapse, making new stars. During the collapse, some of the material will not fall directly onto the central star but, instead, will fall along orbits at various distances from the star. All the orbits will share some motion in one plane, the "equator" of the cloud, and motions in other directions will be reduced or eliminated by collisions, resulting in an orbiting disk.

In 1974, Lynden-Bell and Pringle suggested that the peculiar spectra of a newly-discovered class of young stars, the T Tauri stars (named after the first known star of this type), looked remarkably like the spectra expected for a disk, at least in theory. Astronomers later discovered that many T Tauri stars are expelling symmetrical jets of gas. In 1984, two groups, one led by me, and the other by Gary Grasdalen, obtained an infrared image of a T Tauri star called HL Tau. The area of emission around HL Tau was elongated on the sky, nearly perpendicular to the jets of gas previously found emerging from that star.

These images of disks were observed with the Hubble Space Telescope. A, B, and C are disks seen in silhouette against the Orion nebula. The other three images show disks close to edge-on. Image D, HH 30, shows a jet perpendicular to a dark disk seen edge-on with a gently flared shape. Images E and F show disks around main sequence stars BD+31° 643 and Beta Pictoris.

A

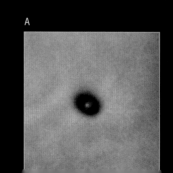

B

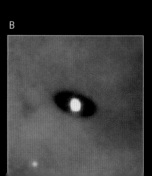

The infrared emission looked like a circumstellar disk seen edge-on. We thought it might be the progenitor of a planetary system similar to the early solar nebula.

Over the next several years, Anneila Sargent and I made observations at radio wavelengths to map the carbon monoxide gas around HL Tau. We found that the gas too was elongated as if in a disk. These maps allowed us to trace the gas velocities using the Doppler effect, and suggested that the gas was trapped in the star's gravitational field. The evidence for a disk surrounding HL Tau appeared to be quite strong.

During this period, NASA launched the Infrared Astronomical Satellite (IRAS), enabling astronomers to study infrared radiation from many young stars. Several groups recognized

that many T Tauri stars emitted excessive energy at long infrared wavelengths, indicating the presence of cool disks.

By 1990, using IRAS data, astronomers had discovered that almost half the T Tauri stars in the dark interstellar clouds of a region called Taurus-Auriga showed evidence of disks; HL Tau was not an isolated anomaly. We found that a similar fraction of stars emitted energy at millimeter wavelengths. This longer wavelength radiation essentially showed that disks of circumstellar dust were the only viable explanations for the data. We were able to derive the masses of the disks for the first time, and we discovered that the great majority had sufficient mass to build planetary systems like our own.

Most specialists in star formation had by then

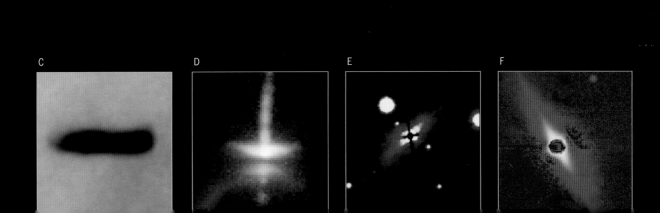

C
D
E
F

accepted disks as the best explanation for the accumulated data on T Tauri stars. But the evidence was largely indirect, and it was often possible to interpret the data with different theories for the distribution of dust near these stars. The first pictures of disks taken with the Hubble Space Telescope demonstrated unequivocally that the dust distributions followed the theoretical pattern of a disk. C. Robert O'Dell and Zheng Wen saw striking examples of disks in silhouette against the Orion Nebula in 1996 (see the essay by C. Robert O'Dell in this section).

Since most of the gas and dust in the disks around other stars is cold, it does not emit visible light. But the cold disk material emits copious amounts of infrared and microwave radiation, at wavelengths longer than can be seen by the human eye. The disks thus became readily detectable about fifteen years ago, when advances in infrared and millimeter-wave astronomy brought them into view. They are relatively easy to see with modern instruments, especially in the infrared spectrum. The infrared radiation from disks is several orders of magnitude stronger than that emitted by the stars at similar wavelengths.

These circumstellar disks typically contain more than ten times the mass of all the planets in our solar system in the form of gas and dust. They have diameters more than twice as large as Pluto's orbit, and are thin enough so that any planets would orbit in a common plane, like that of our own solar system. We see disks around stars a few million years old, so we know the disks persist long enough to build at least small planets. These disks show some observational evidence for particles larger than dust grains and for gaps of the sort we would expect if young planets clear regions in the disks as they orbit the stars. None of this evidence proves that the disks are making planetary systems,

but there is a remarkable correspondence between the predicted properties of planet-forming disks and the observed properties of disks around other stars. Since disks are a natural consequence of star formation, and they provide the opportunity for planet formation, it follows that other stars are likely to have their own planetary systems.

If circumstellar disks do commonly create planetary systems, what does that say about our place in the universe? There are more than 100 billion stars in our Galaxy. If even a few percent of these had disks that evolved to planetary systems like our own, there are roughly a billion such systems in the Milky Way alone. With the Hubble Space Telescope, we can see at least one million galaxies per square degree in the sky, which means there are more than forty billion observable galaxies in the universe. If there are a billion planetary systems per galaxy, the number of potential homes for life is on the order of 10^{18}, a staggeringly large number.

Life appeared rather quickly on Earth, when the conditions allowed it to do so. The first fossils date from within 1.1 billion years of Earth's genesis as measured by the age of the oldest meteorites. But asteroid impacts made the first 800 million years or so of this period a hostile time for life, so there was life on the surface within about 300 million years of the time when it could have survived. That is a short span compared with the age of the Earth itself, about 4.6 billion years, and even shorter compared to the age of the universe, at least 12 billion years by modern estimates.

If life on Earth developed so rapidly, relative to the cosmic timescale, could life itself be an improbable occurrence in the universe? It could, of course, and it may be that we are just the fruit of an incredible series of coincidences. But the prevalence of disks that are the birth sites of planets, and the probability that the number of planets is a few billion billions causes me to believe that there must be some other life out there.

It is a long leap from circumstellar disks to concluding that life is common. But the leap from disks to planets is not so large. None of the newly-discovered planets is enough like the Earth to be suitable for life as we know it, but the prevalence of disks suggests that some planets like the Earth should arise around some stars. It is theoretically easier to construct Earth-like planets than the recently-discovered giant planets around nearby stars. There should thus be many other "Earths" out there. And if there are many Earths, it would be surprising if we were truly alone in the universe.

Circumstellar disk around the star HD 141569, about 320 light-years away. A dark band separates the inner part of the disk from a fainter outer region. The central star is blacked out by the telescope, allowing the faint disk to be seen. Near infrared image from the Hubble Space Telescope.

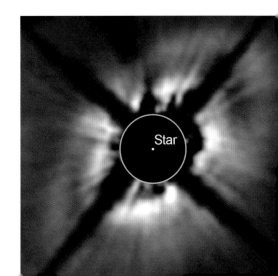

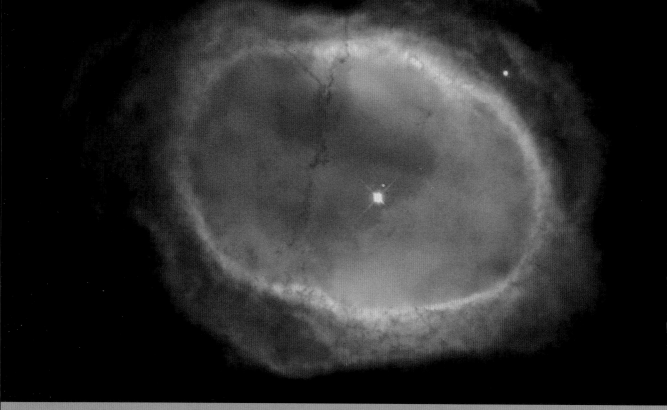

Planetary Nebulae

Adam Frank

In the last decades of the twentieth century, astronomers began to acquire a new understanding of how stars like our Sun age and die. For "solar-type" stars, the end is a long, hard process. But, like the fabled phoenix, from the ashes of these stars arise what are arguably the most beautiful structures in the universe—the gaseous interstellar clouds known as Planetary Nebulae. Using powerful new instruments like the Hubble Space Telescope, astronomers are now piecing together the story of planetary nebulae and of our Sun's ultimate fate.

Adam Frank is an Assistant Professor of Astrophysics at the University of Rochester.

The story of a star's life and death involves two principal players. First there is gravity, pulling the star incessantly inward, crushing its matter toward the center. The second is nuclear fusion, the joining of light elements to create heavier ones. At the center of a star, fusion reactions release the energy that supplies the gas pressure to support a star against gravity. As long as the inward pull of gravity and the outward pressure of fusion energy are balanced, a star stays serene and constant. As a star ages, however, it begins exhausting its nuclear fuel. Gravity gains the upper hand, and the end begins.

In the battle between fusion and gravity, astronomers divide stars into two classes: Massive and Low Mass. The dividing line is fuzzy but currently it rests at approximately eight times the mass of the Sun. Stars with more than eight "solar masses" will fuse a succession of heavier elements together, releasing enough energy to stave off the eventual gravitational crush. But once a massive star begins creating iron, its story is almost over. Iron fusion takes more energy than it gives up. After iron is created, fusion at the core stops. With all support lost, in less than a second the star falls inward in a collapse so violent that the forces generated blow it apart in a titanic supernova explosion. Supernovas are so powerful that we can see them, literally, across the universe.

Like an ill-fortuned Hollywood idol, the brilliant death by supernova of a massive star catches everyone's attention. But supernovas are rare events, simply because massive stars are rare. Most stars, like most people, are ordinary (for stars, ordinary means low mass). But, like the silent dramas accompanying an ordinary person's life, the end of an ordinary star is no less dramatic or enigmatic than that of their blazing massive cousins.

The fireworks accompanying the death of ordinary low mass stars like our Sun may be more muted than a supernova but no less spectacular. As a star like the Sun passes into its final state, it also tears itself apart. Unlike a supernova, however, the process happens over thousands of years. The wreckage left over from this stellar unravelling is visible to us as a glowing cloud of gas and dust called a planetary nebula. Through the eyes of the Hubble Space Telescope, these clouds have been resolved into vast luminescent shapes of astonishing beauty and complexity. Planetary nebulae appear in forms as varied as geodesic spheres, gaseous peanuts, thin jets of plasma, and strangely variegated pinwheels. Interpreting the phenomena embedded in this menagerie of forms poses a challenge for astronomers who must, somehow, extract the science of stellar death from the enigmatic clouds.

The first step in understanding planetary nebulae is to know that they have nothing to do with planets. The name derives from eighteenth-century astronomers who, peering through relatively crude telescopes, discovered objects that appeared neither as points of light like stars nor as distinct disks like planets, but rather as cloudy disks. They christened these objects planetary nebulae, after the Latin word for cloud (nebula).

Planetary nebulae are actually the debris left over as dying stars are ripped apart by their own powerful outward-flowing winds. The shapes of these objects, which span light-years, outline hypersonic shock waves produced by the collision of the winds from one era of the star's evolution with those from the preceding era. Astronomers have recently learned to see planetary nebulae as a kind of gaseous fossil, with the history of the star imprinted in the shape of the cloud. The story extracted from these cosmic boneyards is a tale of high galactic artistry in the midst of incredible violence.

One of the dilemmas astrophysicists face is that they can't make planetary nebulae in their laboratories. The only way to probe the structure and evolution of planetary nebulae is via high-speed supercomputers. The equations describing the collisions of stellar winds are huge, complex, and elaborately interconnected. Trying to solve these equations with pencil, paper, and will power is usually impossible. A supercomputer can solve the equations by taking many tiny steps over and over again, a million times a second. After a few hours or weeks (depending on how super the supercomputer is), the evolution of a model planetary nebula over ten thousand years is captured in the computer's memory for detailed study.

The story the computers reveal is called the "Interacting Stellar Winds" model of planetary nebulae. With this theory astronomers have been able to reproduce, with startling accuracy, the images they see in the sky. In this model, the violent winds that create a planetary nebula are also the engine that turns a bloated red giant star into the burnt-out cinder of a white dwarf, an evolutionary fate common to all low mass stars.

During their middle years, stars like the Sun fuse hydrogen into helium in their cores, liberating nuclear energy that keeps the gas pressure high enough to counteract the weight of gravity. When the hydrogen runs out, the star's core collapses in on itself, contracting until it becomes hot enough to burn its own ashes. The core now fuses helium into the heavier elements carbon and oxygen. But the helium burning floods the star with energy faster than it can radiate away at the surface. The outer layers of the star absorb the excess energy like a sponge, swelling outward in the process. The star then expands into the characteristic distended figure of a red giant.

As the bloated star ages, the extended outer atmosphere cools and contracts, then soaks up more energy from the star and expands yet again. Each repetition of this cycle makes the atmosphere act like a massive piston. These pulsations pump the red giant's atmosphere into space in a dense wind that blows outward at speeds up to thirty kilometers per second. In as little as 10,000 years some red giants lose an entire Sun's worth of matter this way. Eventually this "slow" wind strips the star down close to its fusion core. In a few thousand years, it will be nothing but carbon and oxygen ash, a dead "white dwarf." White dwarfs are the cinders of solar type stars. The compressed carbon ash in a white dwarf makes them something like giant diamonds in the sky. They can support themselves against gravity forever, purely by the pressure of electrons rushing between the crystal carbon lattice.

When the dense wind of the expelled stellar atmosphere first exposes the core, everything changes. While the center of the core is composed of inert carbon ash, the surface is still a hellish mix of fusion reactions among the remaining hydrogen and helium nuclei. The surface layers release a torrent of energetic photons, mostly in the form of ultraviolet light. The photons drive whatever atmosphere is left into space, creating a tenuous high-velocity wind. This "fast" wind, with speeds up to 5,000 kilometers per second, quickly overtakes the slow wind ejected earlier. The two winds collide with the force of a trillion one-megaton H-bombs. That's when the cosmic sculpting and the fireworks really begin.

Figure 1: A sample of the stunning diversity of planetary nebula shapes is amply revealed in these six images. Each of these expanding gas clouds was formed by a solar-type star at the end of its life. In each case, a wind from the central star inflates and shapes a bubble via its interaction with another wind ejected earlier in the star's lifetime. This figure shows that planetary nebulae appear in forms as simple as concentric spheres and as wild as cosmic corkscrews. From left to right, top to bottom, the less-than-poetic names and shapes of these planetary nebulae are as follows: IC 3568, round; NGC 6826, elliptical; NGC 3918, bipolar with jets; Hubble 5, bipolar; NGC 7009, elliptical with jets; NGC 5307, point-symmetric.

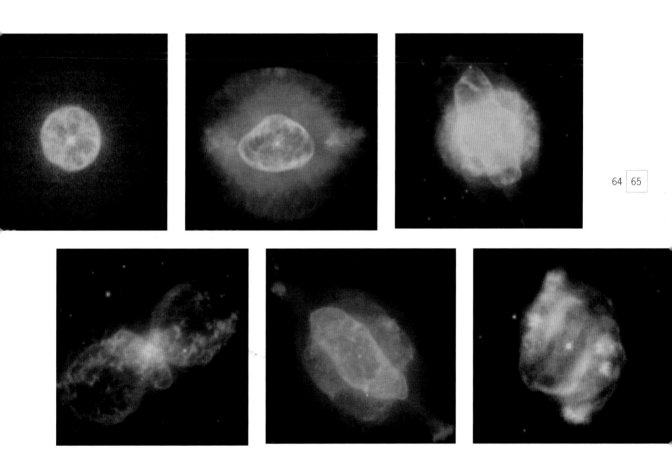

The collision of the fast wind from the exposed core and the previously ejected slow wind creates shock waves. (A shock wave occurs whenever a gas is pushed faster than it can flow out of the way.) As the wave moves through the gas, it violently smashes together the gas molecules like cars in a highway pile-up. When the fast wind slams into the slow wind, a shock wave moves outward, accelerating and compressing the slow gas as it sweeps through it, squeezing it into a dense shell of ions. At the same time another shock wave bounces off the slow wind, traveling back through the fast wind, toward the star. This rebounding shock jerks the fast wind to a near halt. The violent deceleration heats the fast wind to more than 10 million degrees, creating a vast hot bubble of gas. Ultimately, the result is a kind of shock wave layer cake. The inner shock wave is closest to the star, surrounded by the hot bubble, which in turn is surrounded by the dense shell and its outer boundary, the outer shock wave.

The shocks produce light by compressing and heating the gas. The slow wind shell of shocked gas glows most intensely because it is so dense. The gas in the hot bubble is too rarefied to produce much light, in spite of its enormous temperatures. Thus, the shell and the outer shock wave are the glowing forms we see when we view a planetary nebula from the Earth. The observed shape of the planetary nebula depends on the apparent shape of the outer shock. Figure 1 shows some examples.

Round planetary nebulae result from collisions between spherical winds. But most planetary nebulae are not round. Nebulae of other shapes can still be explained by the collision of stellar winds, however, if astronomers assume the slow wind itself is not perfectly symmetrical. This could occur if the amount of mass leaving a star were not the same everywhere on the surface. If, for example, more mass were driven off from the equator of the star than from its poles, the slow wind would assume a flat, disklike shape.

Observations of very young planetary nebulae in fact show such dense "donuts" of gas flowing away from the central star.

Astronomers do not yet know how stars actually make non-spherical winds, but they have a few theories. For example, if the red giant is orbiting a companion as part of a binary star system, then the gravity of the other star might pull the slow wind into the shape of a disk. The outer shock forming behind this flattened cloud could then quickly blow out through the tenuous poles, where relatively little matter would stand in the way. But along the equator the shock would push slowly through the densest parts of the disk. After just a few thousand years, the shock wave layer cake would be distorted into a peanut or elliptical shape, depending on the shape of the slow wind. The more matter spewed out along the equator rather than the poles, the more peanut-shaped, or "bipolar," the final planetary nebula.

While this model of interacting stellar winds has been extremely successful in accounting for the majority of planetary nebula shapes, Nature has shown herself to be a far more inventive sculptor than astronomers. As the Hubble Space Telescope has cataloged an ever-increasing number of planetary nebulae, it has become clear that some other explanation is needed. There is growing evidence that the dizzying variety of shapes cannot all be explained by the interacting winds model. What else is required? One possibility is strong magnetic fields in the stellar winds. The magnetic fields would exert forces in ways that gases alone cannot, perhaps winding the nebulae into the extraordinary pinwheel shapes sometimes observed. While the jury is still out on this possibility, astronomers are clear on one point: the new physics required to explain planetary nebulae will lead to a new view of the stars that created them. Who could have imagined that the death of stars could be so beautiful or so productive?

Friedrich Bessel (1784–1846), German astronomer and mathematician who first measured stellar parallax.

Friedrich Bessel and the Companion of Sirius

Important advances in astronomy often come from increased precision of observations. The work of the German astronomer and mathematician Friedrich Wilhelm Bessel (1784–1846) provides a good illustration. Among his many other accomplishments, Bessel developed techniques to measure the positions of stars with far greater accuracy than previously possible. For example, he made the first precise measurements of refraction of light by the Earth's atmosphere. Refraction is the bending of light rays as they pass at an angle through different substances, like glass, water, or air. When a star is near the horizon, its apparent position can differ from its true position by as much as the diameter of the Moon. This effect was poorly-quantified until Bessel studied it in 1811. His tables of refraction allowed observers to measure

star positions to an unprecedented accuracy of less than a tenth of a second of arc (the size of a small coin as seen from a mile away).

This kind of precision in astrometry (the branch of astronomy that measures stellar positions) allowed Bessel in 1838 to find the first reliable distance to another star. He discovered the long-sought stellar parallax—the extremely tiny shift in the apparent position of a star when observed from opposite sides of the Earth's orbit. Since the size of the Earth's orbit was known, the observed parallax angle allowed Bessel to calculate the distance to the star 61 Cygni by triangulation. This discovery also provided the most convincing proof that the Earth really moves around the Sun.

Bessel went on to obtain precise positional measurements of Sirius, the brightest star in the sky. His observations revealed that Sirius was slowly changing its position as if it were being pulled around in orbit by the gravity of another star. In 1844, Bessel had a sufficient number of precise observations to announce that Sirius must have an unseen companion. The orbital period of the two stars around each other turned out to be about fifty years.

Astronomers searched for the companion star but couldn't find it. Finally, in 1862, after Bessel died, the American telescope maker Alvan Clark, while testing a new telescope on the bright star Sirius, actually discovered the companion. It was indeed a star, but so very faint that it was almost lost in the glare of Sirius. Because the companion was about twice as far as Sirius from their common center of mass, it had to weigh about half as much (like a child twice as far from the center of a see-saw balancing an adult). Why then was the companion almost a thousand times fainter?

Around 1915, Walter Adams at Mt. Wilson Observatory obtained the spectrum of the companion and was astonished to find that the faint star was nearly three times hotter than Sirius. Using the laws of physics, astronomers can calculate the size of a star if they know its temperature and luminosity (light output). To be so hot and yet so faint, the companion of Sirius had to be as small as the Earth, but its mass, calculated from Bessel's astrometry, equalled that of the Sun. Here was a star with the mass of the Sun packed into a volume no larger than the Earth. It had to be about three million times more dense than water. A thimbleful of this stuff would weigh about ten tons on Earth! The companion of Sirius was made of some strange new form of matter, far beyond anything in human experience. The nature of such dense objects, now called white dwarfs, remained a complete mystery until the development of quantum mechanics—the physics of atomic particles.

Sirius, the brightest star in the sky, with its faint white dwarf companion, Sirius B.

In 1930, a young Indian graduate student, Subrahmanyan Chandrasekhar, on a sea voyage to England to study astronomy at Cambridge, applied the new quantum ideas to the physics of stellar structure. He realized that when a star like the Sun exhausts its nuclear fuel, it will collapse due to its own gravity until a new form of pressure comes into play. This pressure is due to the so-called Pauli exclusion principle, which prevents the electrons in matter from getting too close to one another. The fact that the electrons cannot be compressed beyond a certain point determines the very high but stable density of a white dwarf. Chandrasekhar found that electron pressure can support a white dwarf only if the star has less than 1.4 times the mass of the Sun. More massive stars would continue to collapse to some then-unknown fate. This idea was the theoretical breakthrough that pointed the way to neutron stars and black holes, and would later earn Chandrasekhar the Nobel Prize for Physics.

Although difficult to observe due to their small size, white dwarfs actually turn out to be quite common in the Galaxy. In fact, we now know that all low mass stars like the Sun collapse into white dwarfs when they run out of nuclear fuel.

The story of the companion of Sirius has a peculiar sequel. In 1950 the French anthropologist Marcel Griaule published a study of the Dogon tribe of Mali in West Africa, in which he described an elaborate Dogon ceremony centered around the star Sirius. The Dogon informed Griaule that Sirius was accompanied by a very heavy, metallic companion star that was completely invisible. Excited by this information, UFO enthusiasts took it as proof that the Dogon had been visited long ago by aliens who taught them about the white dwarf companion of Sirius. How else could they possibly have known about it? But the astronomer Carl Sagan suggested a much simpler and more human explanation. He noted that the existence of the unseen, dense companion of Sirius was widely known in Europe long before Griaule recorded the Dogon mythology. The tribe had often been visited by missionaries and travelers. It is quite possible, even probable, that one of these visitors, perhaps during an exchange of sky lore with the Dogon, told them about the companion of Sirius, particularly since the Dogon used the appearance of Sirius to mark the changing seasons. The Dogon, recognizing a good story when they heard it, incorporated the invisible companion of Sirius into their own traditions, which were later recorded by Griaule.

When Stars Collide

Michael M. Shara

- Somewhere in the universe, a thousand pairs of stars are colliding right now.

- During stellar encounters, some stars are reborn, while others perish.

- The oldest fossil star clusters are largely governed by collisions between stars.

- We can identify the stars produced by stellar collisions, although no one has ever observed such a collision.

Michael M. Shara is Curator and Chair of the Department of Astrophysics at the American Museum of Natural History.

All of these hard-to-believe statements are true, though they might seem more science-fiction than fact at first glance.

Most of the stars in our Galaxy are enormously far apart compared to their sizes. If you picture a typical star as a basketball, its nearest stellar neighbor would be another basketball on the opposite side of the Earth. Stars in the solar neighborhood typically are moving at about 75,000 kilometers per hour (50,000 mph) with respect to each other, but, because of their staggeringly large separation, they have had almost no chance of colliding during the thirteen billion year history of the Milky Way Galaxy. Other catastrophes will happen to the Sun (and Earth) in the very distant future, but a collision with a star is not one of them.

Why then should we expect stars ever to collide? It turns out that results of the cosmic billiards game are quite different for stars living in dense star clusters. A globular cluster is a collection of up to a million stars crammed into a volume reserved for about a hundred stars in more typical parts of the Galaxy (the solar neighborhood, for example). Trapped by their mutual gravitational pulls, cluster stars move blindly about like bees in a swarm. Given enough time—and 13 billion years is a long enough time—collisions of some of the tightly packed stars become inevitable.

Detailed calculations show that in the densest central regions of some globular clusters, fully half the stars have undergone one or more collisions since the beginning of the universe. Because there are roughly 1,000 billion globular clusters in the cosmos, and a collision typically lasts for a few hours, we find that at any moment (like right now) about 1,000 pairs of stars are undergoing a collision somewhere in the universe. The closest such event is about a billion light-years away, explaining why no astronomer has yet identified a collision in progress.

What happens when two stars collide? The outcome depends on several factors, but most of all on the kinds of stars running into each other. There are seven distinct objects that most astronomers agree are stars, nearly stars, or stellar corpses. In order of decreasing average density, these are: black holes, neutron stars, white dwarfs, brown dwarfs, main sequence dwarfs, giants, and supergiants. The Sun is a main sequence dwarf. Any of these seven types of stars can collide with any of the others, yielding twenty-eight unique pairs of cosmic fender-benders to consider. Fortunately there are a few simple principles that govern what happens during collisions. Most important of these is density contrast. Simply put, the higher density star will suffer much less damage than the lower density star during a collision, just as a high density cannon shell is barely marked as it blows a watermelon to shreds.

The greatest difference among the seven kinds of stars listed above is their average density. A neutron star is 500 trillion times denser than the Sun, and the Sun is 20 million times denser than a supergiant. The Sun is subject to terrible damage and likely destruction in any encounter with a much denser black hole, neutron star or white dwarf, but would suffer only modest changes in most encounters with tenuous giants or supergiants. The outcomes of collisions between similar stars (say, two ordinary sunlike dwarfs or two neutron stars) are determined by the contest between the gravity that binds and kinetic energy that disrupts the system.

Let's consider first a head-on collision between a sunlike star and a vastly more dense white dwarf (the results are similar for a neutron star). While each meanders along at a sedate 50,000 mph or so in isolation, mutual gravitational attraction

70 71

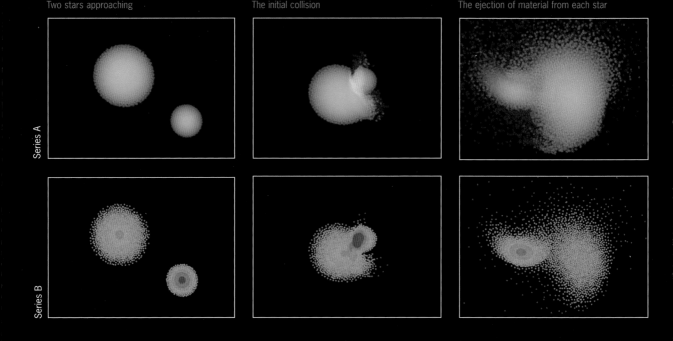

Series A

Series B

Two stars approaching

The initial collision

The ejection of material from each star

Collision of Unequal Mass Stars
Researchers increasingly create computer simulations and visualizations to support theoretical models of astronomical events. The images shown here are stills, taken from a 1,500-frame movie, depicting the collision of unequal mass stars. Time advances from left to right.

Series A shows the whole view of the collision; the visualization represents the event as seen from the exterior of the stars. The smaller star "plows into" the larger one, loses momentum, and eventually merges.

Series B shows a view of the collision from the interior of the stars; the visualization represents the movement of the particles in a slice through the mid-plane of the collision. Color represents gas density. High-density material in the small star (red) remains together as the small star collides with the larger one. The smaller star eventually settles almost intact within the center of the larger one.

Simulations by Joshua Barnes, University of Hawaii.

accelerates the stars to over 1,000,000 mph as they approach each other on a collision course. The white dwarf is about the size of the Earth, and roughly 100 times smaller than its target. It penetrates the sunlike star at hypersonic speed, setting up a massive shock wave that compresses and heats the entire target star above thermonuclear ignition temperatures. Only an hour is required for the white dwarf and its attendant shock wave to smash its way through the hapless sunlike star, but the damage is irreversible. The super-heated sunlike star releases as much fusion energy in that hour as it normally does during 100 million years. The resulting pressure forces the shocked star to expand at speeds far above its escape velocity. Within a few hours the sunlike star literally blows itself apart. Ironically, the source of this catastrophe—the white dwarf—continues blithely on its way, with only its outermost layers warmed by the disrupting sunlike star and the nuclear conflagration it leaves behind.

The ensuing collision and merging | The two stars spiraling together as they merge | The stars merged

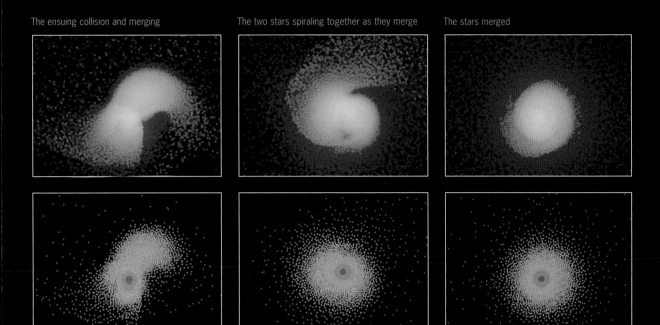

A very different sequence of events occurs when the stars colliding head-on in a typical star cluster are of the same type, density, and size. Only the approach and collision speeds are similar to the previous example. Calculations show that during the collision, as the initially spherical stars increasingly overlap, they compress and distort each other into nearly half-moon shapes. Temperatures and densities never become high enough to ignite the destructive thermonuclear burning accompanying the collision of a white dwarf with a sunlike star. A few percent of the total mass of the two stars is ejected perpendicular to the direction of stellar motion, but then mixing and overlap take over. Within an hour or so, the two incoming stars are mixed together—fused for all time into a single, new star twice as massive as each of its progenitors. This is one of the strongest predictions of stellar collision theory.

It has taken a generation, but the Hubble Space Telescope has given astronomers the tools needed to begin testing that theory. Stars like our Sun are gigantic nuclear furnaces. They fuse hydrogen into helium and heavier elements via nuclear reactions in their cores while their cooler surface hydrogen remains unchanged. The amount of fuel left in the core of a star is like a clock, directly measuring how many millions of years of fusion power remain before the star dies. Isolated stars like our Sun have no way of replenishing the core hydrogen fuel allotment they were born with; hence their life span is preordained. Colliding stars of similar density are fundamentally different. Because there is mixing of surface and core layers during a collision, the coalesced star that emerges has more core hydrogen than either of the stars that made it. This kind of core-hydrogen-rich star thus has its clock reset by the collision.

Determining the age of a star cluster is quite simple: it's just the age of the most luminous hydrogen-fusing star in the cluster. Computer models compared with direct measurements of energy production in bright cluster stars yield stellar ages. Because the clock for a coalesced star is reset by the collision, it is predicted to be unsynchronized with the clocks of all the normal stars in a cluster core.

Astonishingly, this effect has now been detected in the cores of several dozen globular clusters orbiting the Milky Way Galaxy and several neighboring galaxies. Star clusters that are 13 billion years old contain, at their very centers, dozens to hundreds of stars no older than one billion years. These phoenix-like, reborn stars are called "blue stragglers," because they are hotter and bluer than their older, non-coalesced counterparts, and because their youthful characteristics make them appear to straggle behind other more evolved stars. The mass of one blue straggler has been directly measured, and it is almost precisely twice that of all the other stars in its cluster, a strong indication of coalescence. For astronomers, the discovery of blue stragglers has been like finding teenagers living in a retirement home.

Near misses are more common than head-on collisions. Extremely close passages (missing by only one or two star diameters) may result in the mutual gravitational capture of two stars to form a very close, strongly interacting binary pair. These hypothetical objects are being as avidly sought as blue stragglers. Just a few dozen very close binaries contain as much energy in the form of orbital motions as the gravitational energy holding together an entire cluster of a million stars. Each of these binaries can act as a source or sink of energy, flinging passing stars out of the cluster, or more tightly binding them to the cluster core. Thus the evolution of clusters, and their eventual dissolution may be largely driven by near-collisions.

The observational and theoretical pictures I've just sketched are far from complete. Collisions between stars are usually more glancing than head-on, and the results of these encounters are much trickier to predict, though very rapidly spinning stars might result. Collisions involving black holes or giant stars may yield exotic, poorly understood types of stars. Finally, blue stragglers might also be formed when two stars, born very close together and locked in a gravitational embrace for billions of years, spiral into one another and merge into one star. Subtle but potentially measurable differences are predicted for blue stragglers formed via this different mechanism. Determining which formation theory is correct (collision or binary merger or both) will require many more observations and computer models. Only if we're extraordinarily lucky will a direct collision occur close enough (say, within 10 million light-years) within the lifetimes of today's astronomers to be directly observable. I'm convinced, however, that the technologies of astronomers one hundred years from now will allow direct and routine tests of stellar collision theory.

Cecilia Payne and the
Composition of the Stars: Profile

❝ There is no joy more intense than that of coming upon a fact that cannot be understood in terms of currently accepted ideas. ❞

—Cecilia Payne

What are the stars made of? The answer to this fundamental question of astrophysics was discovered in 1925 by Cecilia Payne and explained in her Ph.D. thesis. Payne showed how to decode the complicated spectra of

starlight in order to learn the relative amounts of the chemical elements in the stars. In 1960 the distinguished astronomer Otto Struve referred to this work as "the most brilliant Ph.D. thesis ever written in astronomy."

Cecilia Payne (1900–1979) was born in Wendover, England. After entering Cambridge University she soon knew she wanted to study a science, but was not sure which one. She then chanced to hear the astronomer Arthur Eddington give a public lecture on his recent expedition to observe the 1919 solar eclipse, an observation that proved Einstein's Theory of General Relativity. She later recalled her exhilaration: "The result was a complete transformation of my world picture. When I returned to my room I found that I could write down the lecture word for word." She realized that physics was for her.

Later, when the Cambridge Observatory held an open night for the public, she went and asked the staff so many questions that they fetched "The Professor." She seized the opportunity and told Professor Eddington that she wanted to be an astronomer. He suggested a number of books for her to read, but she had already read them. Eddington then invited her to use the Observatory's library, with access to all the latest astronomical journals. This simple gesture opened the world of astronomical research to her.

England, though, was not in Payne's professional future. She realized early during her Cambridge years that a woman had little chance of advancing beyond a teaching role, and no chance at all of getting an advanced degree. In 1923 she left England for the United States, where she lived the rest of her life. She met Harlow Shapley, the new director of the Harvard College Observatory, who offered her a graduate fellowship.

Harvard had the world's largest archive of stellar spectra on photographic plates. Astronomers obtain such spectra by attaching a spectroscope to a telescope. This instrument spreads starlight out into its "rainbow" of colors, spanning all the wavelengths of visible light. The wavelength increases from the violet to the red end of the spectrum, as the energy of the light decreases. A typical stellar spectrum has many narrow dark gaps where the light at particular wavelengths (or energies) is missing. These gaps are called absorption "lines," and are due to various chemical elements in the

star's atmosphere that absorb the light coming from hotter regions below.

The study of spectra had in fact given rise to the science of astrophysics. In 1859, Gustav Kirchoff and Robert Bunsen in Germany heated various chemical elements and observed the spectra of the light given off by the incandescent gas. They found that each element has its own characteristic set of spectral lines—its uniquely identifying "fingerprint." In 1863, William Huggins in England observed many of these same lines in the spectra of the stars. The visible universe, it turned out, is made of the same chemical elements as those found on Earth.

In principle, it seemed that one might obtain the composition of the stars by comparing their spectral lines to those of known chemical elements observed in laboratory spectra. Astronomers had identified elements like calcium and iron as responsible for some of the most prominent lines, so they naturally assumed that such heavy elements were among the major constituents of the stars. In fact, Henry Norris Russell at Princeton had concluded that if the Earth's crust were heated to the temperature of the Sun, its spectrum would look nearly the same.

When Payne arrived at Harvard, a comprehensive study of stellar spectra had long been underway. Annie Jump Cannon had sorted the spectra of several hundred thousand stars into seven distinct classes. She had devised and ordered the classification scheme, based on differences in the spectral features. Astronomers assumed that the spectral classes represented a sequence of decreasing surface temperatures of the stars, but no one was able to demonstrate this quantitatively.

Cecilia Payne, who studied the new science of quantum physics, knew that the pattern of features in the spectrum of any atom was determined by the configuration of its electrons. She also knew that at high temperatures, one or more electrons are stripped from the atoms, which are then called ions. The Indian physicist M. N. Saha had recently shown how the temperature and pressure in the atmosphere of a star determine the extent to which various atoms are ionized.

Payne began a long project to measure the absorption lines in stellar spectra, and within two years produced a thesis for her doctoral degree, the first awarded for work at Harvard College Observatory. In it, she showed that the wide variation in stellar spectra is due mainly to the different ionization states of the atoms and hence different surface temperatures of the stars, not to different amounts of the elements. She calculated the relative amounts of eighteen elements and showed that the compositions were nearly the same among the different kinds of stars. She discovered, surprisingly, that the Sun and the other stars are composed almost entirely of hydrogen and helium, the two lightest elements. All the heavier elements, like those making up the bulk of the Earth, account for less than two percent of the mass of the stars.

Most of the mass of the visible universe is hydrogen, the lightest element, and not the heavier elements that are more prominent in the spectra of the stars! This was indeed a revolutionary discovery. Shapley sent Payne's thesis to Professor Russell at Princeton, who informed her that the result was "clearly impossible." To protect her career, Payne inserted a statement in her thesis that the calculated abundances of hydrogen and helium were "almost certainly not real."

She then converted her thesis into the book *Stellar Atmospheres*, which was well-received by astronomers. Within a few years it was clear to everyone that her results were both fundamental and correct. Cecilia Payne had showed for the first time how to "read" the surface temperature of any star from its spectrum. She showed that Cannon's ordering of the stellar spectral classes was indeed a sequence of decreasing temperatures and she was able to calculate the temperatures. The so-called Hertzsprung-Russell diagram, a plot of luminosity versus spectral class of the stars, could now be properly interpreted, and it became by far the most powerful analytical tool in stellar astrophysics.

Payne also contributed widely to the physical understanding of variable stars. Much of this work was done in association with the Russian astronomer Sergei Gaposchkin, whom she married in 1934.

From the time she finished her Ph.D. through the 1930s, Payne advised students, conducted research, and lectured—all the usual duties of a professor. Yet, because she was a woman, her only title at Harvard was "technical assistant" to Professor Shapley. Despite being indisputably one of the most brilliant and creative astronomers of the twentieth century, Cecilia Payne was never elected to the elite National Academy of Sciences. But times were beginning to change. In 1956, she was finally made a full professor (the first woman so recognized at Harvard) and chair of the Astronomy Department.

Her fellow astronomers certainly came to appreciate her genius. In 1976, the American Astronomical Society awarded her the prestigious Henry Norris Russell Prize. In her acceptance lecture, she said, "The reward of the young scientist is the emotional thrill of being the first person in the history of the world to see something or to understand something." As much as any astronomer, she had fully experienced that most important of all scientific rewards.

A grazing encounter between two spiral galaxies will contibute to their eventual merger. Light from the smaller galaxy, IC 2163, in the background at right, silhouettes dust lanes in spiral arms of the larger galaxy, NGC 2207. Hubble Space Telescope.

Section Three: Galaxies

Introduction Steven Soter

A typical galaxy contains many billions of stars bound together by gravity. Astronomers broadly classify galaxies according to shape—as spiral, elliptical, and irregular. The most photogenic galaxies are the great spirals—flat disks with luminous arms spiraling out from a central bulge. The stars and clouds in the disk of a spiral galaxy orbit in the same direction, pulled by the mass lying inside their nearly circular orbits. The flattened disk is due to the organized motion of its stars. Our Milky Way is a large spiral galaxy.

The spiral arms are concentrations of bright young stars and interstellar clouds of gas and dust. The clouds are the debris ejected from earlier generations of stars. These vast clouds are stellar nurseries and recycling centers where stars are born in batches. Our Sun, with its system of planets, was born in such a cloud about 4.6 billion years ago. That cloud is long gone, and the sibling stars of the Sun are spread out around the disk of the Milky Way. But other clouds are forming from the debris of dying stars, and the process of stellar regeneration goes on.

When we look at the faint band of the Milky Way arcing across the sky, we are actually looking directly out into the plane of our Galaxy, where most of the stars are concentrated. We cannot see the whole Galaxy, because the dust clouds in the disk obscure our view of its more distant parts. In fact, we know more about the structure of many other galaxies than about our own, because we can see them clearly from outside, when we look above and below the dust filled plane of our Galaxy.

The disk of our Milky Way is about a hundred thousand light-years across and two thousand light-years thick. It contains a few hundred billion orbiting stars. The Sun is about two-thirds of the way from the center and it takes about 240 million years to complete one orbit. Since its birth, our solar system has made only about twenty orbits around the center of the Milky Way.

In contrast to spirals, the elliptical galaxies show relatively little internal structure. Seen from any direction, they typically have an elliptical cross-section. The stars in elliptical galaxies move in random directions around or through the central region, like a swarm of flying insects. Elliptical galaxies lack the clouds of gas and dust from which new stars are made, so their stars are mostly very old. Elliptical galaxies include both the largest and smallest of galaxies.

Galaxies make large targets for mutual collisions. The distance between the Milky Way and the nearest large galaxy, the great spiral in Andromeda, is only about 20 times larger than the size of either galaxy. In contrast, stars make small targets. The neighboring stars within a galaxy are typically separated by tens of millions of stellar diameters. Collisions between the ordinary stars in a galaxy are rare, while collisions between galaxies themselves are fairly common.

When galaxies collide, they may pass right through one another after a billion years. In many cases, they will then fall back together, become entangled, and finally merge in a mutual gravitational embrace. The stars initially glide past each other without colliding, but the clouds of gas and dust in the galaxies pile up violently against each other, often triggering great bursts of star formation. A galaxy that manages to emerge from a collision may be torn apart by the gravitational encounter. Some of the irregular galaxies appear to have resulted from such collisions.

Many large galaxies possess small satellite galaxies. Our Milky Way has two irregular satellites called the Large and Small Magellanic Clouds. Galaxies are also found in larger groups or clusters, bound by gravity. Giant ellipticals, the largest galaxies by far, tend to live near the centers of dense clusters. They probably "dine" on smaller galaxies that fall into their gravitational grasp, and have grown unusually massive by such "galactic cannibalism."

Galaxies evolve through time. Because the speed of light is finite, we see distant galaxies not as they are now, but as they were long ago, when their light was emitted. The more distant a galaxy, the further back in the past we observe it. Astronomers use telescopes like "time machines," to directly observe what distant galaxies looked like billions of years ago. The evidence from remote regions suggests that there were fewer spiral and elliptical galaxies and more small and irregular ones in the early universe than today. It appears that large galaxies have indeed grown at the expense of smaller ones.

Most of the mass in galaxies has never been directly observed. We know this so-called "dark matter" exists only because we can observe the effects of its gravity on the visible stars. For example, in the Milky Way and other spiral galaxies, the stars in the outer parts of the disk orbit around the center so rapidly that they must be attracted by much more mass than that of the visible stars. Astronomers have evidence that the dark matter extends far beyond the volume occupied by the visible stars of a galaxy. It appears that galaxies are enveloped in enormous spherical "halos" of dark matter. The nature of the dark matter, which constitutes perhaps ninety percent of the mass of the universe, is one of the major unsolved mysteries of astronomy.

One class of theories assumes that dark matter consists of exotic subatomic particles that are invisible because they do not interact with any kind of light. Physicists are searching for such particles using sensitive detectors on Earth. A second class of theories assumes that dark matter consists of ordinary matter in the form of objects of planetary or stellar mass that are too small or non-luminous to see. These bodies might be detected when they pass in front of stars, causing a transient increase in the starlight by the so-called gravitational lens effect.

Astronomers recognized only recently that very small and faint galaxies far outnumber all the others in the universe. Since the so-called dwarf galaxies also contain dark matter, their large population could make a substantial contribution to the total mass of the universe. Additional

Spiral galaxy Messier 104, the "Sombrero", seen nearly edge-on, with its dust lanes sihouetted against the bright central bulge. European Southern Observatory.

dark matter exists in the so-called "low surface brightness" galaxies, which can be as large as the Milky Way but are also extremely faint and hard to observe.

Some of the farthest galaxies evidently produce stupendous amounts of energy from very small volumes in their cores. In some cases, the energy output from these so-called "active galaxies" is seen to fluctuate in as little as a few hours. No source of light can change its total brightness in less time than light takes to go across it. So the rapid brightness variations in such active galaxies means that the energy source must be only a few light-hours across. That's only about the size of our planetary system. Yet the energy output appears to be thousands of times greater than that of an entire normal galaxy.

No ordinary sources can produce such concentrated and enormous amounts of energy. The evidence suggests that supermassive black holes, containing as much as a billion solar masses, lie at the heart of active galaxies. The energy is generated by collisions and friction within a rapidly rotating "accretion disk" of matter as it spirals into the black hole. When the available "fuel" is used up, the active galaxy "turns off," leaving a quiescent supermassive black hole. In fact, astronomers find extremely high stellar velocities in the cores of many normal galaxies, including our own Milky Way, suggesting that supermassive black holes are hiding there as well.

To explore galaxies, we pose the following questions:

How do galaxies change over time?

Henry C. Ferguson, Associate Astronomer at the Space Telescope Science Institute, examines the evidence from the most distant galaxies that show what galaxies were like at earlier epochs in the past.

How can we detect "dark matter"?

Christopher Stubbs, is a faculty member of both departments of Astronomy and of Physics at the University of Washington in Seattle, describes current attempts to pin down dark matter in our Galaxy by observing its influence on the light from background stars.

What are "faint galaxies" and why do they matter?

Julianne Dalcanton, Assistant Professor of Astronomy at the University of Washington in Seattle, shows that faint galaxies far outnumber the rest, and may make a substantial contribution to the total mass of the universe.

What are quasars?
What are supermassive black holes?

Roeland P. van der Marel, an astronomer at the Space Telescope Science Institute in Baltimore, traces the observational evidence that many galaxies went through an early stage of extremely intense energy production, perhaps fueled by the gravity of supermassive black holes.

The spiral galaxy, NGC 1232, is about 200,000 light-years across and twice the size of our Milky Way Galaxy. Its central bulge contains older relatively cool red stars, while the spiral arms have hot young blue stars and active star-forming regions. European Southern Observatory.

Galaxies Through Space and Time

Henry C. Ferguson

One of the major discoveries of the twentieth century was Edwin Hubble's realization in 1929 that the universe is expanding. If we imagine running the observed expansion backwards, we come to a time when the material of the galaxies was all crowded together. It seems therefore that the universe had a beginning and that its age is measureable.

Because light has a finite speed (186,000 miles per second, or one light-year per year), astronomers can literally observe the past. A galaxy 10 million light-years away appears not as it is

Henry C. Ferguson is an Associate Astronomer at the Space Telescope Science Institute, and Deputy Project Scientist for the Next Generation Space Telescope.

Figure 1: Two typical spiral galaxies, NGC 2997 and NGC 891, one seen face-on, the other seen edge-on. The central "bulge" region is typically more yellow than the outer regions, which means that it contains older stars. Star formation is concentrated in the outer parts where there is much dust and gas.

today but as it was 10 million years ago, when the light we now see began its journey. But the technology of Hubble's time could only provide detailed views of galaxies out to a few hundred million light-years away. The corresponding light travel time is much less than the lifetime of a typical star, which is measured in billions of years. Thus the galaxies Hubble could study and classify were all more or less contemporaries in time with our own Milky Way Galaxy. But galaxies at increasing distances represent successively earlier stages in the history of the universe and reveal evidence of long-term evolution through time. The ability to observe that evidence is a recent development.

Let us begin our survey close to home. Our Milky Way Galaxy, which contains about 100 billion stars, stretches about 100,000 light-years from edge to edge. We can see clusters of stars within it that are much younger than the Sun, and others that are much older. Although the age of an individual star is difficult to determine, astronomers can calculate the age of a star cluster based on the characteristic distribution of brightness and color among its member stars. Careful measurements of these quantities show that the Milky Way is about 12 billion years old—more than twice the age of the Sun (4.6 billion years).

The Milky Way is surrounded by a small retinue of "dwarf" galaxies containing anywhere from a few million to a billion stars (see the essay by Julianne Dalcanton in this section). Some of these galaxies appear to be as old as the oldest stars in the Milky Way; others have stars being formed even today.

The nearest big galaxy similar in size to the Milky Way is the Andromeda Galaxy, which can be seen (just barely) with the naked eye on a dark night. It is 2.5 million light-years away, so the light now arriving from Andromeda began its journey when our ancestors were evolving from anthropoid apes.

Both Andromeda and the Milky Way are spiral galaxies. They are relatively flat, with a bright reddish bulge in the center and a striking pinwheel shape in the outer regions, similar to those seen in Figure 1. These spiral arms are typically filled with gas and dust and lots of young stars. Looking around the nearby universe, we also find quite a few football-shaped "elliptical galaxies." These tend to have very little dust and gas, and usually have no sign of ongoing star formation. They evidently formed their stars a long time ago, and probably much more rapidly than spiral galaxies.

Elliptical galaxies tend to be found in clusters. The nearest big cluster of galaxies is in the constellation of Virgo (Figure 2). It contains about 1,000 or so galaxies bright enough to see with our largest telescopes, and probably a lot more that are too faint to see. The Virgo Cluster is 60 million light-years away. Looking at it takes us back in time only 60 million years, still not very far relative to the ages of stars. To see galaxies as they were forming, we have to look back billions of years.

For a long time, looking back that far seemed impossible because such distant galaxies appear extremely faint. Twenty years ago, even heroic nightlong time-exposures with the largest telescopes on Earth (with mirrors sixteen feet across) could record galaxies no fainter than about twenty-third magnitude, or about 6 million times fainter than we can see with our unaided eyes. But a galaxy like the Milky Way would look still fainter if it were more than four billion light-years away. Light from that distance dates back almost to the age of the Sun, but is nowhere near as old as the first stars, which formed about 12 billion years ago.

A lot has changed in the last twenty years. Sensitive charge-coupled devices, which can detect single photons of light, have replaced photographic plates for observing the faint images of distant galaxies. Telescopes have gotten bigger (with mirrors now thirty feet across) and observatories have been carefully engineered to minimize disturbances of the atmosphere around the telescope. The Hubble Space Telescope, unimpeded by the atmosphere, has been providing images with five to ten times the clarity possible from the ground. It is now possible to find galaxies back at a time when the universe was only one or two billion years old.

Measuring the precise distance to a galaxy is extremely difficult. Galaxies come in different sizes and shapes and there is no easy way to tell if a faint galaxy looks faint because it is distant or because it is small. Astronomers generally determine the distance of a remote galaxy by measuring the spectrum of its light.

The farther away a galaxy is, the more time its light has traveled across space to reach us. But the universe is expanding. So the more time the light has traveled, the more its waves have been stretched out by the expansion of space itself. The stretching of visible light waves makes them appear redder, so this effect is called the cosmic "redshift."

Astronomers use the redshift in the spectrum of a galaxy, denoted by the letter "z," as a rough measure of its distance. Galaxies relatively nearby have very low redshifts. The Virgo Cluster, for example, has a redshift $z = 0.005$. This means the wavelengths of light from the stars in that cluster have been stretched by 0.5 percent compared with those emitted by the same atoms measured in the laboratory. By 1990 the deepest surveys with the largest telescopes were able to find galaxies out to redshifts $z = 0.8$, which correspond to "look-back times" of about 7 billion years.

Figure 2: Galaxies in the Virgo Cluster. The two brightest members are giant elliptical galaxies, while spirals and more irregular galaxies can be seen around them.

At these distances, the correspondence between redshift, which we can measure directly, and lookback time starts to become quite uncertain. That is because it depends on the large-scale curvature of space, which we do not know very well. According to Einstein's General Theory of Relativity, the universe is curved by the gravitational attraction of all the matter and energy within it. The curvature is in principle measurable, which is one of the reasons for finding and studying distant galaxies. (If galaxies did not evolve, simply counting how many there are at various distances would be enough to measure the curvature of the universe.) Since we do not know the curvature, we do not know precisely the ages and distances of galaxies with high redshifts. Very roughly, according to a standard theoretical model, if the universe is 13 billion years old today, then it was 5 billion years old at redshift $z = 1$, and 2 billion years old at $z = 3$, and about 1 billion years old at $z = 5$.

In 1995, a major advance was made when the Hubble Telescope spent ten days observing one small patch of sky to get the sharpest image of a random set of distant galaxies. In the same year, astronomers using the Keck Observatory in Hawaii began systematically identifying galaxies at redshifts greater than 3. The Keck astronomers used colors of the galaxies to find which ones of the hundreds of galaxies in their images were likely to be at high redshift. The color selection technique relies on the fact that hydrogen, the most abundant element in the universe, absorbs light below a certain wavelength known as the "Lyman break" in the ultraviolet part of the spectrum. When this spectral feature gets redshifted to longer wavelengths, into the visible part of the spectrum, those distant galaxies have a charac-teristic color that makes them easy to identify.

The Hubble Deep Field image revealed about 3,000 galaxies in a tiny patch of sky the size of a grain of sand held at arms length (Figure 3). The faintest of these galaxies had brightnesses equivalent to one photon per week striking the human eye. While a few looked like the familiar spiral and elliptical galaxies we see nearby, many of them looked more peculiar, with signs that they are either colliding with their neighbors or have not yet settled into a stable structure. Many of the galaxies in the image are so small that astronomers currently believe they are dwarf galaxies.

It is possible that the familiar spiral and elliptical galaxies we see nearby were built up over time by the successive merging of smaller galaxies. Indeed this is precisely what the most popular theories of galaxy formation predict. In these "hierarchical clustering" theories, small enhancements in the density of dark matter in the universe drew in matter around them and contracted under the force of gravity. The gas in these little lumps cooled and formed stars. Over time, many of these dwarf-galaxy-sized lumps collided and merged to form bigger galaxies. If the collision was violent, material got mixed around and ended up looking like an elliptical galaxy. If the collision was less violent, or if gas was slowly drawn in after the collision, a spiral galaxy could form.

In this picture, the process of galaxy formation is gradual. Star formation in galaxies does not turn on all at once, but builds up gradually to a peak and then tapers back off again. By measuring the brightnesses and approximate distances in the Hubble Deep Field images, it has been possible to confirm this general behavior.

Another prediction of galaxy formation theory is that the elliptical galaxies and the bulges of spiral galaxies should form relatively late, many of them at redshifts less than 1 (which

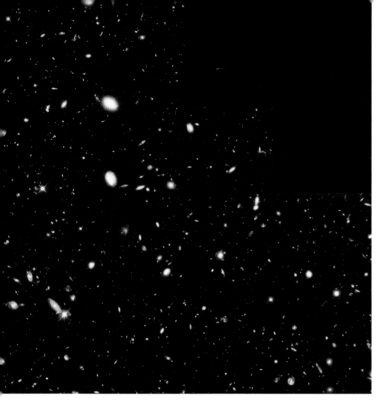

forming stars at rates of hundreds to thousands of times faster than in a mature galaxy like ours, so even if they were doing all that within the first two billion years after the Big Bang, we ought to be able to detect them.

We do not see them. Or more precisely, the objects that we do see at lookback times of 10 billion years do not appear to be forming stars that fast, and do not in general have the sizes or shapes of elliptical galaxies. The galaxies at redshift $z = 3$ and greater that have been detected by the Keck and Hubble telescopes appear to be forming stars at rates only about ten times faster than in galaxies today (see Figure 4).

So either the elliptical galaxies and spiral bulges formed later, or formed their stars more slowly, or are somehow being hidden from us at high redshift. One possibility, which is being actively explored, is that the first few generations of massive stars shed so much material that the galaxies quickly (within about a hundred million years or so) disappeared behind a veil of dust. If that is the case, then we might be able to find them by searching with infrared or radio telescopes, because dust absorbs the ultraviolet and visible light from stars and reradiates it at longer wavelengths. To search for such first generation stars, astronomers have made observations of very deep space with the Infrared Space Observatory, and with a small high-frequency radio antenna on Mauna Kea in Hawaii. In the Hubble Deep Field, a handful of radio sources have been found, but the low resolution of the images from these telescopes makes it difficult to tell which (if any) of the galaxies in the optical images correspond to the source.

On the other hand, the numbers of objects we detect at high redshift and their brightnesses

corresponds to within about the last 8 billion years). This is a rather controversial prediction.

Measurements of the ages of stars in the Milky Way bulge indicate that most of them are about 12 billion years old. Studies of nearby elliptical galaxies give ambiguous answers for the ages, and there is a lot of debate about whether the observations require stars younger than about 5 billion years, or whether most elliptical galaxies are much older than that. So the observations of nearby galaxies do not tell us clearly whether this theory is right or wrong.

The recent data from Keck and Hubble allow us to look back in time and try to find the progenitors of present-day elliptical galaxies. They seem to be there out to redshifts $z = 1$ (about 8 billion years ago) in about the same numbers as we see locally. Beyond that, we see very few, but they would be extremely hard to find if they are already old (and hence relatively dim) at those redshifts. However, if elliptical galaxies formed all their stars within a billion years, they ought to be easy to find when they were undergoing that bright starburst phase. They would be are generally consistent with the hierarchical clustering theories. Given a galaxy at one spot on the sky, the theory can predict the likelihood of finding a neighboring galaxy within a certain distance of it. One of the remarkable findings of the last several years is that predictions of this kind of theory match the measurements of the clustering of Lyman break galaxies extremely well, provided that the Lyman break galaxies are the progenitors of the giant elliptical galaxies we see today in places like the Virgo Cluster.

So perhaps we have seen the progenitors of giant elliptical galaxies, or perhaps not. The debate rages on. As for spirals like the Milky Way, we see them also out to redshifts of z = 1, but beyond that we find few examples. It is still an open question whether their bulges formed via mergers or via a rapid collapse of a gas cloud at high redshift, or through some other process. It is also still uncertain when the disks around these bulges started to grow. The images taken so far provide clues but no hard answers.

We are quite sure we are not seeing all there is to see. The scheduled installation of new cameras on the Hubble Space Telescope will increase its sensitivity and field of view immensely. Then, in 2008 or so, the Next Generation Space Telescope will be launched to peer even closer to the Big Bang and to detect objects that are fainter or are obscured by dust (see the essay by Alan Dressler in Section Six). With these facilities and others coming along we can hope to gradually piece together a consistent picture of how galaxies formed.

Figure 4: The relative rate of star formation in the universe over time, as derived from the Hubble Deep Field and other surveys. The horizontal axis shows the observed redshifts (along the bottom) and the calculated equivalent lookback times (along the top). The latter is based on a theoretical model for a universe 13 billion years old. The vertical axis shows the average rate of star formation, relative to that at present (at zero redshift). The error bars show the uncertainties in the measurements of star formation rate and redshift.

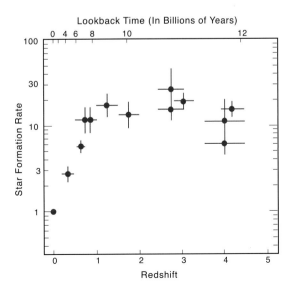

Searching for Dark Matter in Galaxies with Gravitational Lensing

Christopher Stubbs

You Mean Galaxies Aren't Just Visible Stars and Gas?

It comes as something of a surprise that the luminous stars and gas that we can see and photograph are, in fact, the minority component of a typical galaxy. We now know that the beautiful textbook photographs of spiral galaxies are not accurate renditions of all that is there. Most of the Galaxy is made up of a substance that we cannot see and do not yet understand.

Christopher Stubbs is a faculty member of both the Departments of Astronomy and Physics at the University of Washington in Seattle.

How can astrophysicists make this bold claim? Until recently, any determination of the inventory of matter in a galaxy was based on counting the objects in the galaxy that either emit or absorb light. Lately, however, astronomers have developed techniques that allow them to measure directly the mass of a galaxy. By using stars as tracers of the strength of the gravitational pull of a galaxy, we can measure the galaxy's total mass.

The mass of a typical galaxy inferred in this fashion exceeds the mass of its visible stars and gas by as much as a factor of ten. The rest, the galactic "dark matter," is therefore by far the gravitationally dominant constituent of galaxies, including our own Milky Way. Understanding this dark matter is the key to resolving many of the open questions in cosmology and astrophysics.

The Dark Matter Puzzle, or "What's the Matter?"

The evidence for dark matter in galaxies started to accumulate in the mid-1970s. By the following decade, it became clear that all galaxies, including ours, are surrounded by extensive "halos" of dark matter. Just as astronomers used the orbits of the planets about the solar system to calculate the mass of the Sun, they can use the orbital motions of stars to trace the gravitational strength of an entire galaxy. Measurements of the internal motions of many hundreds of galaxies provide incontrovertible evidence that stars, gas, and dust alone cannot account for their observed properties. Dark matter halos are thought to be roughly spherical, extending far beyond a galaxy's stellar component. Just how far these dark matter halos extend, and therefore how much total mass they contain, is a topic of considerable current debate.

Wimps versus Machos

There are two broad categories of dark matter candidates: massive but dark astrophysical objects, and invisible elementary particles. One class of elementary particle candidates for dark matter is called WIMPs, for Weakly Interacting Massive Particles. This play on words reflects the assumption that these particles interact with matter through the "electroweak" force, one of the fundamental forces of particle physics. Not to be outdone, the astrophysical community has dubbed its favorite class of massive dark matter candidates MACHOs, for Massive Compact Halo Objects. Machos are ordinary objects of planetary or stellar mass that are too faint to be seen directly.

Searching for dark matter is difficult. The only evidence we have for its existence comes from its gravitational influence on its surroundings. As far as we know, wimps neither emit nor absorb electromagnetic radiation, which precludes direct detection with the traditional tools of astronomy.

Despite these difficulties, a number of experiments are under way to search for particular dark matter candidates. One class of experiments searches for evidence of rare interactions between wimps and a sensitive detector material on Earth. As wimp interaction rates are expected to be in the range of a few detection events per kilogram of target material per day, the main experimental challenge is to understand and overcome naturally occurring sources of radioactive background "noise." These can either mask, or masquerade as a detection of, dark matter. Wimp experiments with sufficient sensitivity and background discrimination are now moving from the prototype stage to full-scale operation, and we can look forward to important results in the years to come.

Using a Gravitational Lens to Search for Dark Matter in the Milky Way

Perhaps the most dramatic progress in dark matter searches in recent years has been in

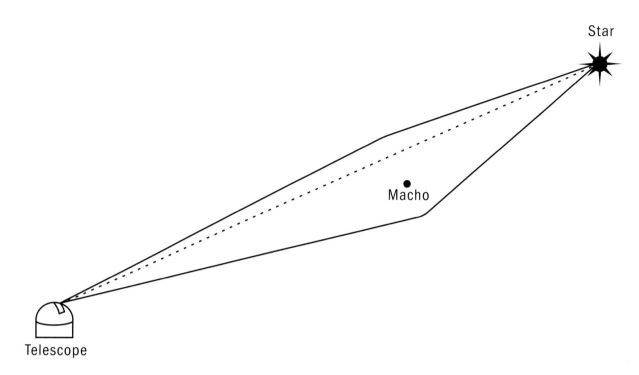

Star

Macho

Telescope

searches for machos, the astrophysical dark matter candidates. These searches exploit the one thing we know for certain about dark matter—that it exerts a gravitational force on its surroundings.

If the dark matter halo of the Milky Way contains drifting machos, occasionally one of them will pass close to the line of sight between us and a more distant star. The light coming from the background star will then be deflected, due to the gravitational warping of space caused by the intervening macho, as predicted by Einstein's Theory of General Relativity. This provides an elegant and effective technique to search for machos in the dark matter halo of the Galaxy; we can look or their gravitational effect on light from more distant stars.

The deflection of the incoming light by the gravitational field of the macho is the same as if an optical lens of astronomical proportions had been placed between us and the background

Schematic representation of gravitational microlensing. Light traveling from a distant star is deflected by a macho (an unseen object of planetary or stellar mass), which acts as a gravitational lens. The star appears to undergo a transient brightening as a result of the lensing.

star, making the star appear brighter. As this type of gravitational lensing is due to stellar or planetary masses, rather than galactic masses, it is termed "gravitational microlensing."

The observable signature of gravitational microlensing relies on the fact that this precise alignment between us, the intervening macho, and a background star is fleeting. The participants in this alignment are in relative motion. Our line of sight to any star sweeps through space due to the Earth's motion within the Galaxy, and we expect any machos in the dark matter halo to have speeds of hundreds of kilometers per hour, just like other halo objects. The signature of microlensing should therefore be a transient brightening of a background star, with a specific shape predicted by the Theory of General Relativity.

The duration of a microlensing event by a halo object with the mass of the planet Jupiter is expected to be about three days. The microlensing technique is therefore well-suited to searching for machos, by making repeated nightly observations of stars using astronomical telescopes.

The effect of microlensing can be substantial, with the apparent brightness of a star temporarily increasing by factors of 40 or more. The duration of a microlensing event depends on the mass of the machos, its location, and its speed across the sky. A low mass macho moving slowly can produce a signal that is indistinguishable from that of a more massive object moving rapidly.

So much for the good news. While the signature of microlensing is unique, and the technique is sensitive to a broad range of masses, the expected rate of microlensing events is somewhat sobering. Even if the dark matter halo of our Galaxy were entirely accounted for by machos, only about one star in a million would appear significantly brighter due to microlensing at any given time.

Microlensing Has Been Detected

At the time of this writing, the number of observed candidate microlensing events exceeds 400. The overwhelming majority of these events are seen towards the Galactic center, and result from microlensing by ordinary stars in the disk of the Galaxy.

The field of microlensing is progressing rapidly. Since the first candidate events were reported in 1993, we have progressed to the stage where certain exotic microlensing events can be interpreted. For example, during a microlensing event that lasts many tens to hundreds of days,

the Earth's orbit around the Sun produces a slight skew in the observed light curve (a plot of the star's apparent brightness with time). This subtle effect is caused by the variation of the Earth's motion relative to the lens over time. We can use it to determine whether the lensing object is nearby or far away, depending on its mass.

Microlensing Towards the Magellanic Clouds—Testing the MACHO Hypothesis

The overwhelming majority of microlensing events so far observed have lines of sight that pass through our Galaxy's disk and towards its center. But a handful of events have been seen towards the Magellanic Clouds (two small satellite galaxies located well above the plane of the Milky Way), and they may provide the key to understanding our Galaxy's dark matter halo. Since microlensing events observed towards the Magellanic Clouds would be dominated not by ordinary stars but by machos in our Galaxy's halo, this is the most sensitive way to detect them.

Over the course of two years of observations of stars in the Large Magellanic Cloud (LMC), our MACHO team has produced the most stringent results to date on the possibility that massive objects, rather than elementary particles, might account for the dark matter halo of our Galaxy. We have carried out over a thousand observations of nearly ten million stars in the LMC. This data set has led to two major dark matter results, described below.

No Low Mass Machos

It is often the case in science that tremendous progress is made by eliminating possibilities. The search for dark matter is no exception. In our LMC data set, we have searched for microlensing events with a duration ranging from a few hours to 200 days. We have seen no LMC events that last less than twenty days. Using the connection between the mass of the lensing object and the duration of the lensing event, we can exclude objects having masses between one-millionth and one-tenth of the mass of the Sun as candidates for the dark matter halo of the Galaxy. Jupiter-mass objects and brown dwarfs (objects intermediate in mass between planets and stars) had been the favored candidates for machos, but they fall squarely within this excluded range. They were thought to be prime candidates for dark matter, because they are not massive enough for their internal pressure to ignite the nuclear burning that makes stars shine. But the microlensing results show that there are not enough of them to account for the dark matter. Excluding such a broad range of dark matter candidates, with high statistical confidence, is a major step forward in dark matter research.

An Excess of Long Duration Events— Detection of the Galactic Dark Matter?

We did, however, detect eight microlensing events towards the LMC, when only about one event was expected from lensing by known stellar populations. These events typically last eighty days, corresponding to machos with a few tenths of a solar mass, although the uncertainty in the mass is quite large.

Is this the long-sought galactic dark matter? Based on models of the structure of the Galaxy and its dark halo that were popular before the microlensing results were announced, the answer would seem to be yes. Taken at face value, the event rate corresponds to our having detected a population of objects that account for at least half the Galaxy's dark matter. Should this prove to be the case, one of the major astrophysical and cosmological puzzles will have been solved.

There are, however, other possible interpretations of the observed microlensing event rate that do not invoke the dark matter halo. The statistical uncertainties are still fairly large, particularly when we are dealing with only a few detected events. There are several possibilities that might account for the excess microlensing events with ordinary stars rather than machos:

- Lensing by foreground LMC stars
- Lensing by some foreground dwarf galaxy or stellar debris from the LMC
- Lensing by some previously unknown extended population of stars from our own Galaxy.

These are testable hypotheses, and we must distinguish between these alternatives and the dark matter interpretation of the observations. We are now laying the groundwork for a next generation microlensing search to resolve unambiguously whether the microlensing is due to massive compact halo objects (dark matter), or to ordinary stars in unexpected places.

There are two possible answers to the experimental question of whether the excess event rate we have seen is due to machos. Either outcome will make this a crucial and successful experiment. If the results support the macho hypothesis, then one piece of the dark matter puzzle will be firmly in place. On the other hand, if further intense scrutiny shows that the detected signal is due to some previously unrecognized population of ordinary stars, then microlensing searches will have essentially eliminated massive objects as viable dark matter candidates. But this result would also be a great step forward in unraveling the mystery of dark matter. We will gain important knowledge in either case.

Vera Rubin, American astronomer who established the presence of dark matter in galaxies, measures spectra in the 1970s.

Vera Rubin and Dark Matter: **Profile**

" In a spiral galaxy, the ratio of dark-to-light matter is about a factor of ten. That's probably a good number for the ratio of our ignorance-to-knowledge. We're out of kindergarten, but only in about third grade. **"**

—Vera Rubin

At age ten, Vera Rubin was fascinated by the stars, as she watched the night sky revolve from her north-facing bedroom in Washington D.C. Although her father was dubious about the career opportunities in astronomy, he supported her interest by helping her build her own telescope and going with her to amateur astronomers' meetings. She got a scholarship to the prestigious women's college Vassar, where she graduated as the only astronomy major in 1948. Applying to graduate schools, Rubin was told that "Princeton does not accept women" in the astronomy program. (That policy was not abandoned until 1975.) Undaunted, Rubin applied to Cornell, where she studied physics under Philip Morrison, Richard Feynman, and Hans Bethe. She then went on to Georgetown University, where she earned her Ph.D. in 1954 (under George Gamow, who was nearby at George Washington University).

After teaching for a few years at Georgetown, she took a research position at the Carnegie Institution in Washington, which had a modest astronomy program. Her work focused on observations of the dynamics of galaxies. She teamed up with Kent Ford, an astronomer who had developed an extremely sensitive spectrometer.

Rubin and Ford used the spectrometer to spread out the spectrum of light coming from the stars in different parts of spiral galaxies. The stars in the disk of a galaxy move in roughly circular orbits around the center. If the disk is inclined to our line of sight, the stars on one side approach us while those on the other side move away. When a source of light moves toward us, we see a decrease in the wavelengths of the light (a shift toward the blue end of the spectrum), and when the source moves away, we see an increase in the wavelengths (a shift toward the red end). This is called the Doppler effect, and the wavelength shift is

proportional to the speed of the light source relative to the observer. Rubin and Ford made careful measurements of Doppler shifts across the disks of several galaxies. They could then calculate the orbital speeds of the stars in different parts of those galaxies.

Because the core region of a spiral galaxy has the highest concentration of visible stars, astronomers assumed that most of the mass and hence gravity of a galaxy would also be concentrated toward its center. In that case, the farther a star is from the center, the slower its expected orbital speed. Similarly, in our solar system, the outer planets move more slowly around the Sun than the inner ones. By observing how the orbital speed of stars depends on their distance from the center of a galaxy, astronomers, in principle, could calculate how the mass is distributed throughout the galaxy.

When Rubin and Ford began making Doppler observations of the orbital speeds in spiral galaxies, they immediately discovered something entirely unexpected. The stars far from the centers of galaxies, in the sparsely populated outer regions, were moving just as fast as those closer in. This was odd, because the visible mass of a galaxy does not have enough gravity to hold such rapidly moving stars in orbit. It followed that there had to be a tremendous amount of unseen matter in the outer regions of galaxies where the visible stars are relatively few. Rubin and Ford went on to study some sixty spiral galaxies and always found the same thing. "What you see in a spiral galaxy," Rubin concluded, "is not what you get."

Her calculations showed that galaxies must contain about ten times as much "dark" mass as can be accounted for by the visible stars. In short, at least ninety percent of the mass in galaxies, and therefore in the observable universe, is invisible and unidentified. Then

Rubin remembered something she learned as a graduate student about earlier evidence for unseen mass in the universe. In 1933, Fritz Zwicky had analyzed the Doppler velocities of whole galaxies within the Coma cluster (see the Profile on Fritz Zwicky in Section Four). He found that the individual galaxies within the cluster are moving so fast that they would escape if the cluster were held together only by the gravity of its visible mass. Since the cluster shows no signs of flying apart, it must contain a preponderance of "dark matter"—about ten times more than the visible matter—to bind it together. Zwicky's conclusion was correct, but his colleagues had been skeptical. Rubin realized that she had discovered compelling evidence for Zwicky's dark matter. Most of the mass of the universe is indeed hidden from our view.

Many astronomers were initially reluctant to accept this conclusion. But the observations were so unambiguous and the interpretation so straightforward that they soon realized Rubin had to be right. The luminous stars are only the visible tracers of a much larger mass that makes up a galaxy. The stars occupy only the inner regions of an enormous spherical "halo" of unseen dark matter that comprises most of a galaxy's mass. Perhaps there are even major accumulations of dark matter in the vast spaces between galaxies, without any visible stars to trace their presence. But if so, they would be very difficult to observe.

And just what is this "dark matter," so far unobserved except by the effect of its gravity on the stars? The question is one of the major unsolved mysteries of astronomy today. Many theoretical and observational astronomers are hard at work trying to answer it.

Vera Rubin continues to explore the galaxies. In 1992, she discovered a galaxy (NGC 4550) in which half the stars in the disk are orbiting in one direction and half in the opposite direction, with both systems intermingled! Perhaps this resulted from the merging of two galaxies rotating in opposite directions. Rubin has since found several other cases of similarly bizarre behavior. More recently, she and her colleagues found that half the galaxies in the great Virgo cluster show signs of disturbances due to close gravitational encounters with other galaxies.

In recognition of her achievements, Vera Rubin was elected to the National Academy of Sciences and in 1993 was awarded the National Medal of Science. But throughout her career, Rubin has not sought status or acclaim. Rather, her goal has been the personal satisfaction of scientific discovery. "We have peered into a new world," she wrote, "and have seen that it is more mysterious and more complex than we had imagined. Still more mysteries of the universe remain hidden. Their discovery awaits the adventurous scientists of the future. I like it this way."

96 | 97

The spiral galaxy ESO 510-13 is about the size of our Milky Way. The warping of its disk may have been produced by a close gravitational encounter with another galaxy. European Southern Observatory.

Faint Galaxies

Julianne Dalcanton

Our Milky Way is a spiral galaxy containing a few hundred billion stars in a disk about 100 thousand light-years across. Since the galaxies most often pictured in popular media have comparably impressive dimensions, it's easy to imagine that such magnificent structures are typical of galaxies. But they are actually far outnumbered by dwarf galaxies. While all galaxies are enormous compared with the tangible world of our daily concerns, astronomers have recently learned to appreciate that most of them are actually minuscule compared to our Milky Way.

Julianne Dalcanton is an Assistant Professor of Astronomy at the University of Washington in Seattle.

One way that we can judge the size of a galaxy is by measuring the amount of light emitted by its constituent stars. The brighter a galaxy, the more stars it must contain. The brightest galaxies we observe live in the centers of clusters, which are the densest accumulations of galaxies in the universe. These galaxies are over a million times brighter than the nearly invisible faint galaxies that we have found swirling around our own Milky Way. To bring home the astounding range of galaxy sizes, imagine that the biggest of all galaxies were the size of a blue whale. On that scale, our own Milky Way Galaxy would then be about the size of a hippopotamus, but the smallest galaxies would be the size of single-celled organisms, visible only through a microscope!

The smallest, faintest members of the galactic zoo are known as dwarf galaxies. Like bacteria, these easily overlooked systems are by far the most numerous of the galaxies. They pervade the nooks and crannies of our local galactic neighborhood, outnumbering more spectacular galaxies like the Milky Way by more than twenty to one. We still don't know exactly how numerous they are. In spite of their proximity, dwarf galaxies are so faint and so low in surface brightness that astronomers are adding more of them to the local census every year. Unlike bright galaxies, which are often marked by easily recognized spiral patterns of stars, dwarf galaxies appear as formless smudges of light. As a result, they are easily overlooked by even the most highly trained, keen-eyed observer. So there are probably vast numbers undiscovered even today.

Why is it important to know the true population of dwarf galaxies? By studying the motions of individual stars within dwarf galaxies, astronomers have become increasingly convinced that these otherwise unimpressive systems must contain far more mass than can be accounted for by the stars alone. The stars within these galaxies are zipping about more rapidly than can be explained by the gravity of only the visible stars. They must be accelerated by the gravitational pull of unseen matter within the galaxy. Astronomers had earlier deduced the existence of this so-called "dark matter" in bright spiral galaxies, where its mass is estimated to be two to five time greater than that of the visible stellar matter. In the faintest dwarf galaxies, however, astronomers are finding that the dark matter accounts for a factor of fifty times more than the mass in stars! Coupled with the large numbers of dwarf galaxies, the measurements of such a large proportion of dark matter may suggest that dwarfs carry a far larger fraction of the mass of the universe than was previously realized.

For decades, all galaxies with diffuse features having low surface brightness were assumed to be small and unimportant detritus scattered throughout the universe. In the last decade, however, astronomers have discovered that some of the objects that appear to be dwarf galaxies are in fact "imposters." Instead, this newly-discovered class of galaxies, while sharing with dwarf galaxies a diffuse appearance and a high content of dark matter, are in fact as large as the Milky Way. These galaxies, known as "low surface brightness galaxies" or LSBs, are far more massive than their dwarf counterparts studied nearby. Thus they have an even higher chance of being an important contributor to the mass of the universe. Unfortunately, in spite of their larger mass and size, LSBs remain just as difficult to find as dwarf galaxies, and their true numbers remain poorly known to this day.

Even with the astounding differences among galaxies, they all share one important quality: they are evolving, dynamic systems. If nature

The Pegasus Dwarf Galaxy, a member of our Local
Group of galaxies is almost hidden in the glare of bright
foreground stars in our own Milky Way. This dwarf
is only about 2,000 light-years across, or about two
percent the size of our Milky Way.

abhors a vacuum, it truly adores change. Initially, galaxies were thought to be static forms hanging in an unchanging firmament. However, with the widespread acceptance of the Big Bang, astronomers soon understood that galaxies must have had a beginning, which means that they evolved and could not always have looked as they do today. This notion has been reinforced by observations of the birth and death of stars, which we can see in our own galaxy and close neighbors, and by evidence for the evolution of galaxies, which we have been able to observe through space and time (see the article by Henry C. Ferguson in this section).

One of the true surprises of the last two decades is how much of the change seen in the nearby part of the universe is due to the faintest galaxies. With their unimpressive, diffuse appearance, such dwarf galaxies were historically assumed to be old, inactive, and quiescent. Instead, detailed analysis of the "fossil record" contained within their stellar populations suggests a complicated, chaotic history that we do not yet understand.

It seems that many of the smallest galaxies went through multiple violent bursts of star formation. Apparently, these galaxies would lie dormant for billions of years and then suddenly commence star formation at a tremendous rate, only to slip back into hibernation again. We still don't know what triggers these bursts. We have often assumed that some close interaction between the galaxy and a more massive neighbor is necessary to initiate these episodes of star formation. However, some galaxies with this chaotic history are very isolated, and are unlikely to have been corrupted by the influence of a close neighbor. Another possibility is that faint galaxies with more sedate histories are much harder to find. Consisting only of older, decaying stars, the more quiescent faint galaxies may simply have faded beyond our ability to detect them.

Given the complex history revealed within nearby dwarfs, we must ask how such tiny galaxies could effectively hold onto the raw material needed to fuel these multiple bursts of star formation. The violence of these star formation episodes should create powerful winds capable of expelling all the gaseous star-making material from the galaxy. In spite of their high dark matter content, dwarf galaxies still have too little mass to provide enough gravitational pull to resist the onslaught of stellar winds and supernovas. Yet, these galaxies do indeed survive this violent process, and somehow spring back to life after a period of dormancy.

The rich star formation behavior which we see in our nearby dwarf companions may mirror the behavior seen among all normal galaxies in the distant past. Spectra of distant galaxies often reveal the characteristic signatures of the same star formation bursts experienced more recently by the nearby dwarf galaxies. These faraway glimpses into our past suggest that brighter galaxies may share the intermittent star formation history of dwarf galaxies. While we struggle to understand the mechanisms that control star formation in individual nearby dwarfs, which are close enough to examine in occasionally excruciating detail, we may also be elucidating the very mechanisms that have regulated star formation in all galaxies throughout time.

Again and again, the larger cosmological importance of dwarf galaxies stands in sharp contrast to their otherwise unimpressive appearance and size. Overlooked and often unappreciated, the smallest galaxies have important lessons for our understanding of the universe as a whole. While difficult to find, dwarfs are typically so close that they reveal many of the secrets hidden in larger but more distant galaxies. They tell us complicated tales of a rich universe of change, with hidden matter in unlikely places. These fascinating stories have truly just begun to be told.

Active Galaxies and Black Holes

Roeland P. van der Marel

Galaxies are the primary building blocks of the universe. Our Sun resides in a rather flat galaxy that reveals itself as a luminous band across the sky: the Milky Way. Few other galaxies are visible with the naked eye, but even modest telescopes reveal that thousands of other galaxies occupy our nearby corner of the universe. With state-of-the-art telescopes, astronomers have been able to detect and study galaxies at the other side of the universe. A century of intensive study of galaxies near and far, young and old, has now produced a deep understanding of the structure and evolution of galaxies.

Roeland P. van der Marel is an astronomer at the Space Telescope Science Institute in Baltimore.

Possibly the most remarkable finding has been that exceptionally energetic phenomena occur in the centers of a few percent of all galaxies. This is revealed by unusually strong emission across the electromagnetic spectrum of light, including radio waves, X-rays, and gamma rays. These "active galaxies" are believed to be powered by the appetite of massive black holes in their centers, which gobble up matter through their relentless gravitational pull and transfer mass to energy in the process. Has the existence of these black holes been proven? What are their masses? Where did they come from? How common are black holes in galaxies? What are the details of the mechanisms that generate the observed activity? What is the relation between active and normal galaxies? With an ever-increasing range of powerful telescopes, astronomers have recently started to unveil the answers to many of these intriguing questions.

The cataloging of nearby galaxies, or nebulae, as they were then known, started in the eighteenth century with the work of Charles Messier. Until the 1920s, it remained unanswered whether these were gaseous clouds within our own galaxy, or separate "island universes" (i.e., galaxies like our own). Even at that time several cases had already been observed of what later came to be known as galaxy activity, including bright gaseous emission from the centers of some spiral galaxies and a luminous jet in galaxy number 87 of Messier's catalog (M87 in the constellation Virgo). Not surprisingly, these observations were viewed as mere curiosities in a time when the nature of galaxies themselves remained to be established.

As galaxies gradually came to be understood as collections of stars like our Milky Way, the peculiar properties of some galaxies gained interest. In the 1940s and 1950s, research into active galaxies seriously took off. This started with Carl Seyfert's spectroscopic work on the centers of spiral galaxies, and culminated with the beginning of radio astronomy. Pioneering work by Karl Jansky and Grote Reber had demonstrated that some astronomical objects emit radio waves. After improvements in radio observing techniques brought on by the Second World War, astronomers realized that some of the most energetic natural radio sources were galaxies. These galaxies appeared normal on photographs made with visible light—they looked no different from galaxies without radio emission. The radio emission of the active galaxies was puzzling because it could not be attributed to a collection of stars and gas under normal conditions.

The mystery deepened with the discovery of quasars in the 1960s. These radio sources correspond to point-like objects on visible light photographs. Their light is unusually shifted towards redder wavelengths, which indicates that they are very distant. To be visible to us so far away, they must be very luminous. It was eventually realized that quasars are active galaxies that appear point-like because the active nucleus outshines all the stars in the surrounding galaxy. Only recently has it become possible to detect the underlying starlight in quasars. After the initial discovery of quasars, it was found that they do not necessarily have to be strong radio sources. We now know that they can be either "radio-loud" or "radio-quiet." Modern theories try to explain this difference as a result of the three-dimensional orientation of the active galaxy with respect to our line of sight.

There are many astronomical sources of X-rays, a highly energetic form of light. Fortunately for life on Earth, X-rays cannot penetrate our atmosphere. The systematic exploration of X-ray

The Hubble Space Telescope, by making observations from above the Earth's atmosphere, has given astronomers an unprecedented view of the matter surrounding black holes that reside in the centers of active galaxies.

emission from astronomical objects therefore had to await the development of satellites. As X-ray astronomy came to maturity in the 1970s and 1980s, astronomers found that active galaxies were among the brightest sources of X-ray emission in the sky. In fact, strong X-ray emission is probably the single most unifying characteristic of all active galaxies. Satellites have shown that some active galaxies are also strong emitters of gamma rays, an even more energetic form of light.

In active galaxies, a region near the center produces enormous amounts of emission not just in X-rays and visible light but across the entire electromagnetic spectrum. The emission is often variable, from which it can be deduced that the emitting region must be

very small. Nuclear fusion, the energy source known to power stars (and hydrogen bombs), is not efficient enough to explain the total energy output from active galaxies, and astronomers have been able to provide only one plausible alternative: accretion of matter onto a super-massive black hole.

A black hole is an object that is so massive, yet so small, that neither matter nor light can escape its gravitational pull. Matter near a black hole will fall into it, collide with other matter spiraling in, and heat up while doing so. The energy that it emits as a result is responsible for the light observed in the vicinity of a black hole. To explain the observed amounts of energy, the black holes in active galaxies must have masses of a million to a billion times that of the Sun.

Already at the end of the eighteenth century, John Michell in England and Pierre-Simon Laplace in France had hypothesized the existence of objects from which light could not escape. Just before his death in the First World War, the German astrophysicist Karl Schwarzschild found the solution to Albert Einstein's equations of General Relativity for the case of a black hole. Nonetheless, black holes were long viewed more as a mathematical curiosity rather than a physical reality. This changed in the 1960s when they became the favored explanation for active galaxies. The term "black hole" was coined by John A. Wheeler in 1968, and has since made a lasting impression. Nonetheless, for many years it remained to be proven that active galaxies do indeed contain black holes.

The most direct way to detect the presence of a black hole in a galaxy is through its gravitational pull. Matter far from the black hole will move primarily under the gravitational influence of the stars and the (as yet unidentified) "dark matter" in the galaxy, which make up most of the mass. However, close to the center of the galaxy, the gravity from the black hole dominates. Stars and gas in that region move much faster than they would if there were no black hole. Supermassive black holes can therefore be detected by observing rapidly moving stars or gas near the center of a galaxy, provided that the relatively small region where the black hole gravity dominates can be distinguished by the observations.

The spatial resolution for optical observations from Earth is limited by turbulence in our atmosphere. The Hubble Space Telescope circumvents this limitation and delivers observations that are up to ten times sharper than from the ground. It has therefore provided a breakthrough in the detection of black holes in galaxy centers. There are now approximately twenty galaxies for which the presence of a supermassive black hole has been established.

The mass of each black hole is roughly proportional to the mass of the galaxy itself, and makes up approximately half a percent of the total galaxy mass.

The activity in a galaxy ceases when the black hole runs out of "fuel," or when this fuel stops being efficiently transformed into light. Hence, a normal galaxy may have had an active phase in the past, and if so would still have a massive black hole lurking in its center. To test this idea, astronomers have used the Hubble telescope not only to study the centers of active galaxies but also of normal galaxies. Indeed, black holes are found in normal galaxies as well and may even exist in the centers of all galaxies. In our own Milky Way Galaxy, ground-based observations detected rapidly moving stars. These are believed to move under the influence

A dust disk 3,700 light-years across in the center of the active galaxy NGC 7052, as seen by the Hubble Space Telescope. Spectroscopic observations show that the disk is spinning rapidly around a 300 million solar-mass black hole in the galaxy center.

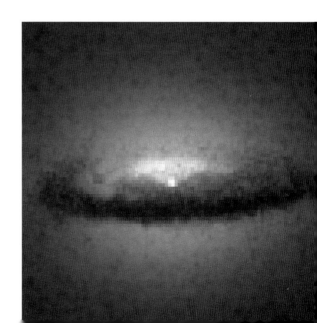

of the gravitational pull of a black hole with three million times the mass of our Sun. Black holes have also been detected in Messier 31, the neighboring giant spiral galaxy in Andromeda (just visible to the naked eye), as well as in its smaller companion Messier 32. Neither of these galaxies is active.

The light from distant active galaxies and quasars has taken billions of years to reach us and provides information from a time when the universe was young. Observations show that quasars were much more numerous in the past than they are now. The number of quasars peaked when the universe was about ten percent of its current age, about one billion years after the Big Bang. The number of quasars in the early universe and their energy output provide information on the total mass that must have accumulated into black holes through quasar activity. The results are consistent with the black hole masses that are being found with the Hubble telescope in nearby galaxies. Hence, it is clear what happened to the population of luminous quasars that lit up the universe when it was young—these quasars have turned into the normal galaxies that we see around us today.

Although we are beginning to understand the evolutionary relation between normal and active galaxies, many questions remain unanswered. How do the black holes in galaxies form? Why do bigger galaxies grow bigger black holes? Do black holes form before or after the galaxies? In 2009, NASA plans to launch its Next Generation Space Telescope, which will focus on the formation and evolution of galaxies and will provide more insight into these issues. NASA's new X-ray satellite Chandra is expected to provide improved information on the energy generation processes in active galaxies. X-ray observations of emission from iron atoms near black holes may determine for the first time whether the black holes are rotating. The spin of a black hole plays a role in the production of the jets that are often seen in radio observations. Also, galaxies sometimes collide and merge, and if they contain black holes, these will merge as well. This causes massive ripples, called gravitational waves, in the space-time fabric of the universe. These may be observed for the first time with the Laser Interferometer Gravitational-Wave Observatory (LIGO), or with the Laser Interferometer Space Antenna (LISA) satellites to be launched by the European Space Agency. Either way, there is little doubt that the future will have many more exciting discoveries in store on the subjects of black holes and active galaxies.

John Michell and Black Holes

A black hole is a volume of space where gravity is so strong that nothing, not even light, can escape from it. This astonishing idea was first announced in 1783 by John Michell, an English country parson. Although he was one of the most brilliant and original scientists of his time, Michell remains virtually unknown today, in part because he did little to develop and promote his own path-breaking ideas.

Michell was born in 1724 and studied at Cambridge University, where he later taught Hebrew, Greek, mathematics, and geology. No portrait of Michell exists, but he was described as "a little short man, of black complexion, and fat." He became rector of Thornhill, near Leeds, where he did most of his important work. Michell had numerous scientific visitors at Leeds, including Benjamin Franklin, the chemist Joseph Priestley (who

discovered oxygen), and the physicist Henry Cavendish (who discovered hydrogen).

The range of his scientific achievements is impressive. In 1750, Michell showed that the magnetic force exerted by each pole of a magnet decreases with the square of the distance. After the catastrophic Lisbon earthquake of 1755, he wrote a book that helped establish seismology as a science. Michell suggested that earthquakes spread out as waves through the solid Earth and are related to the offsets in geological strata now called faults. This work earned him election in 1760 to the Royal Society, an organization of leading scientists.

Michell conceived the experiment and built the apparatus to measure the force of gravity between two objects of known mass.

VII. *On the Means of discovering the Distance, Magnitude, &c. of the Fixed Stars, in consequence of the Diminution of the Velocity of their Light, in case such a Diminution should be found to take place in any of them, and such other Data should be procured from Observations, as would be farther necessary for that Purpose. By the Rev.* John Michell, *B. D. F. R. S. In a Letter to* Henry Cavendish, *Esq. F. R. S. and A. S.*

Read November 27, 1783.

29. If there should really exist in nature any bodies, whose density is not less than that of the sun, and whose diameters are more than 500 times the diameter of the sun, since their light could not arrive at us; or if there should exist any other bodies of a somewhat smaller size, which are not naturally luminous; of the existence of bodies under either of these circumstances, we could have no information from sight; yet, if any other luminous bodies should happen to revolve about them we might still perhaps from the motions of these revolving bodies infer the existence of the central ones with some degree of probability, as this might afford a clue to some of the apparent irregularities of the revolving bodies, which would not be easily explicable on any other hypothesis; but as the consequences of such a supposition are very obvious, and the consideration of them somewhat beside my present purpose, I shall not prosecute them any farther.

Title and excerpt from John Michell's 1783 paper which first described the concept of a black hole. The paper appeared in *Philosophical Transactions of the Royal Society of London.*

Cavendish, who actually carried out the experiment after Michell's death, gave him full credit for the idea. The measurement yeilded a fundamental physical quantity called the gravitational constant, which calibrates the absolute strength of the force of gravity everywhere in the universe. Using the measured value of the constant, Cavendish was able for the first time to calculate the mass and the average density of the Earth.

Michell was also the first to apply the new mathematics of statistics to astronomy. By studying how the stars are distributed on the sky, he showed that many more stars appear as pairs or groups than could be accounted for by random alignments. He argued that these were real systems of double or multiple stars bound together by their mutual gravity. This was the first evidence for the existence of physical associations of stars.

But perhaps Michell's most far-sighted accomplishment was to imagine the existence of black holes. The idea came to him in 1783 while considering a hypothetical method to determine the mass of a star. Michell accepted Newton's theory that light consists of small material particles. He reasoned that such particles, emerging from the surface of a star, would have their speed reduced by the star's gravitational pull, just like projectiles fired upward from the Earth. By measuring the reduction in the speed of the light from a given star, he thought it might be possible to calculate the star's mass.

Michell asked himself how large this effect could be. He knew that any projectile must move faster than a certain critical speed to escape from a star's gravitational embrace. This "escape velocity" depends only on the size and mass of the star. What would happen if a star's gravity were so strong that its escape velocity

exceeded the speed of light? Michell realized that the light would have to fall back to the surface. He knew the approximate speed of light, which Ole Roemer had found in the previous century (see his Profile in Section Four). So it was easy for Michell to calculate that the escape velocity would exceed the speed of light on a star more than 500 times the size of the Sun, assuming the same average density. Light cannot escape from such a body, which would, therefore, be invisible to the outside world. Today we would call it a black hole.

Michell got the right answer, although he was wrong about one point. We now know, from Einstein's relativity theory of 1905, that light moves through space at a constant speed, regardless of the local strength of gravity. So Michell's proposal to find the mass of a star by measuring the speed of its light would not have worked. But he was correct in pointing out that any object must be invisible if its escape velocity exceeds the speed of light. This concept was so far ahead of its time that it made little impression.

The idea of black holes was rediscovered in 1916, after Einstein published his theory of gravity. Karl Schwarzschild then solved Einstein's equations for the case of a black hole, which he envisioned as a spherical volume of warped space surrounding a concentrated mass and completely invisible to the outside world. Work by Robert Oppenheimer and others then led to the idea that such an object might be formed by the collapse of a massive star. The term "black hole" was itself coined in 1968 by the Princeton physicist John Wheeler, who worked out further details of a black hole's properties.

The most common black holes are probably formed by the collapse of massive stars. Larger black holes are thought to be formed by the sudden collapse or gradual accretion of the mass of millions or billions of stars. Most galaxies, including our own Milky Way, probably contain such supermassive black holes at their centers.

Astrophysical theory allows black holes to come in many sizes, and the size of a black hole is simply proportional to its mass. Thus, a black hole with the mass of the Earth would be about an inch across, one with the mass of the Sun would be a few miles across, and one with the total mass of the Milky Way Galaxy would be about a light-year across. The larger a black hole, the lower its average density, and it is conceivable that our entire observable universe is a supermassive black hole within a larger universe.

Michell suggested that we might detect invisible black holes if some of them had luminous stars revolving around them. In fact, this is one method used by astronomers today to infer the existence of black holes. We have observed numerous systems in which matter, whether gas clouds or entire stars, is moving so fast that only the concentrated mass of a black hole could be responsible for it.

While black holes strongly influence the space immediately around them, the notion that they behave like cosmic vacuum cleaners, sweeping up everything in the neighborhood, is a popular fallacy. If the Sun were somehow collapsed to form a black hole, the orbital motion of the planets would be unaffected. The central mass would remain the same, so the planets would feel the same gravity as before. What distinguishes a stellar black hole is its very small size and high density. This allows other bodies to get very close to the center of mass, where the gravity is extremely intense. But it does not increase the pull of gravity far away from the mass.

When John Mitchell concieved of black holes in 1783, very few scientists in the world were mentally equipped to understand what he was talking about. It is not surprising that the concept sank into complete obscurity and had to be rediscovered in the twentieth century. ☄

Section Four: Universe

The large scale structure of the universe, on the scale of
hundreds of millions of light years, shows clusters and
superclusters of galaxies arrayed in a network of filaments.

Introduction Steven Soter

The universe is all the matter, energy, space, and time that exists. We can observe only a part of it. The rest lies beyond our cosmic horizon, which is about 13 billion light-years away in all directions. When astronomers talk about the universe, they usually mean the observable universe. The entire universe may be infinite.

The observable universe contains about 100 billion galaxies. Many of them are found in groups and clusters, and these in turn are arrayed in superclusters of galaxies. Our own Milky Way and some two dozen other galaxies form the so-called Local Group, which is a small part of the Virgo Supercluster. This supercluster contains a few thousand galaxies and spans about 150 million light-years, or roughly one percent the radius of the observable universe.

The superclusters are arrayed in a tangled web of filaments and sheets, leaving large volumes relatively devoid of galaxies. This pattern, called "bubbles and voids," seems to characterize the large-scale structure of the universe.

The universe is expanding. Actually, space itself is expanding and carrying the clusters of galaxies along for the ride, like bubbles entrained in a flowing stream. The clusters are not shooting through space like bullets. Rather, they are nearly at rest in space, but the "fabric" of the space between them is expanding. To visualize this, imagine that an infinitely elastic rubber band represents one dimension of space, and that coins attached to it at various points represent clusters of galaxies. If you then pull the ends of the band apart at a constant speed, the "space" between the "clusters" increases. With uniform expansion, the speed of separation of any two clusters is proportional to the distance between them.

Increase the number of dimensions by one, and the band becomes a rubber sheet stretched in two dimensions. Increase the number again, and the sheet becomes a uniformly expanding volume in three dimensions. As the volume of space expands, the coins embedded in it all move apart from each other, but they do not themselves expand. This represents the fact that in the real universe, galaxies and clusters retain their sizes as space expands because their gravity is strong enough to hold them together. The expansion of space becomes apparent only at the level of superclusters and the large-scale structure.

Light arriving from atoms in distant galaxies has longer wavelengths than light emitted by the same kinds of atoms in the laboratory. Since red light occupies the long wavelength end of the visible spectrum, this effect is called the cosmic redshift. In 1929, Edwin Hubble showed that the farther away a galaxy, the greater the redshift of its observed light. Astronomers assume that this effect is due to the expansion of space, which stretches out the waves of light as they travel through it. The farther light travels through expanding space, the greater its redshift. We can thus use the redshift of a remote galaxy to estimate its distance.

If space is expanding, then the clusters of galaxies were closer together in the past. If we imagine running the expansion backwards in time, then at some remote epoch, everything in the observable universe must have been packed together at enormous density. According to the prevailing theory, the universe was born at that time in an explosion of space called the "Big Bang," which launched the expansion that continues to this day.

If the universe always expanded at the same rate as now, then we could simply calculate its age as the time required for the clusters of galaxies to reach their present separation. But the rate of expansion has not been constant. The mutual gravity of everything in the universe must have slowed down the expansion in the past. And as the galaxies moved farther apart, this force of attraction must have weakened, according to the law of gravity. At the same time, a mysterious long-range repulsive force appears to be at work to speed up the expansion. As the attractive force of gravity gets weaker, the repulsive force increasingly comes to dominate. In fact, it appears as though the expansion of the universe may be accelerating.

The observations only loosely constrain the theoretical models that describe the interplay of these large-scale forces. As a result, cosmologists calculate different values for the age of the universe, with estimates generally falling between 12 and 15 billion years. This accounts for the different ages cited by various authors in this volume. To simplify the following discussion, we will assume that the universe is about 13 billion years old.

The universe that emerged from the Big Bang would have been extremely hot. As it expanded and cooled, energy condensed into particles of matter, including protons and neutrons. Some of them fused to produce the light element helium. After about five minutes, the temperature had fallen too low to sustain further nuclear fusion.

In the early universe, a swarm of free electrons efficiently scattered the light, so that space was everywhere like an opaque yet glowing fog. But the electrons lost energy as the universe expanded and cooled. By about 500,000 years after the Big Bang, the temperature of space had fallen to about three thousand degrees above absolute zero, comparable to the surface of a red giant star. The electrons were by then moving slowly enough to be captured by protons, allowing the first neutral atoms to form. Because bound electrons no longer scatter light, the universe became transparent. Henceforth, light would carry images across space.

Since that event, the observable universe has expanded by a factor of about a thousand. This produced a thousand-fold stretching of the light waves originally emitted by the glowing fog, shifting them from the visible to the microwave part of the spectrum. Microwave radio telescopes still detect this redshifted light— the faint remnant glow of the Big Bang—as the Cosmic Microwave Background Radiation, coming from all directions in the sky. The temperature of space, measured by the microwave background radiation, is now about three degrees above absolute zero.

Because light travels at a finite speed (300,000 kilometers per second), the farther out in space we look, the further back in time we see. We can never see a distant galaxy as it is now, but only as it was long ago, when it emitted the light now arriving. Remote galaxies should look younger than those in our neighborhood. And because the universe is expanding, the most remote regions of space should also appear more crowded with galaxies than the nearby ones.

As we look ever farther out into space, we should see the universe as it looked before the birth of the first stars and galaxies. But at even greater distances, the faint remnant glow from the Big Bang looms up to obscure our view. We cannot see what the universe looked like before about 500,000 years after the Big Bang, because it was then opaque. The Big

World Picture (what we see)

Figure 1: Our "World Picture," or the universe as we observe it. In this schematic view, the Milky Way Galaxy is at the center of the observable universe. Because light travels at a finite speed, the farther out we look in space, the further back we see in time. In our immediate cosmic neighborhood we see mature galaxies like our own, which contain older stars and planets. Farther away in space and time, we see younger galaxies, and beyond them protogalaxies and quasars. Finally, at a distance of about 13 billion light-years, we see the cosmic microwave background radiation, the remnant glow of the Big Bang. Hidden just beyond this glowing surface lies our cosmic horizon. Because the universe is expanding, in the past the galaxies should appear closer together than in our neighborhood. An observer anywhere else in the universe would see approximately the same thing.

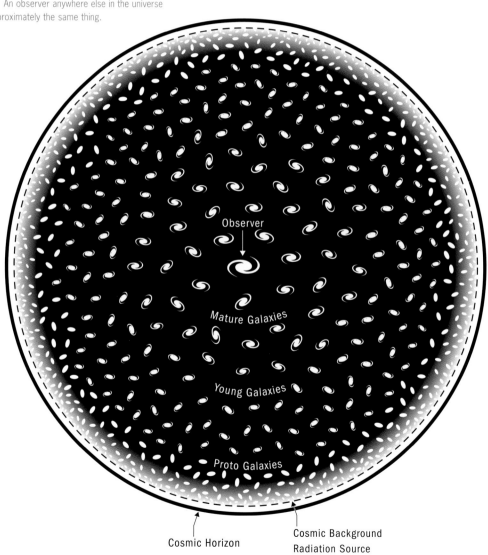

Observer

Mature Galaxies

Young Galaxies

Proto Galaxies

Cosmic Horizon

Cosmic Background Radiation Source

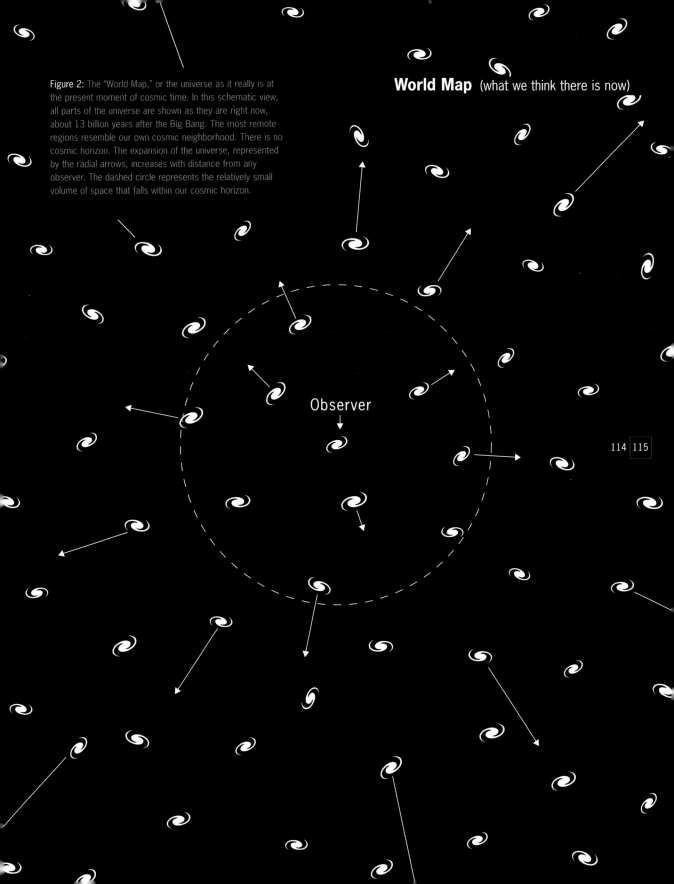

Figure 2: The "World Map," or the universe as it really is at the present moment of cosmic time. In this schematic view, all parts of the universe are shown as they are right now, about 13 billion years after the Big Bang. The most remote regions resemble our own cosmic neighborhood. There is no cosmic horizon. The expansion of the universe, represented by the radial arrows, increases with distance from any observer. The dashed circle represents the relatively small volume of space that falls within our cosmic horizon.

World Map (what we think there is now)

Observer

Bang itself lies hidden just behind the screen of the background radiation, in the same way that the interior of the Sun lies hidden below the opaque photosphere of its surface.

The observable universe extends in every direction to about 13 billion light-years, the distance light has traversed in the age of the universe. We do not know what lies beyond that cosmic horizon. More remote galaxies probably exist there but we cannot see them since there has not been enough time in the age of the universe for their light to reach us. If we could wait another billion years, our cosmic horizon would recede to 14 billion light-years, and some of those galaxies would then lie within our observable universe.

Figure 1 is a schematic two-dimensional view of the universe as we actually observe it. We are at the center of a spherical volume which extends to our cosmic horizon. Space and time as we observe them are inextricably linked: we can only see remote regions as they looked in the past. In every direction, the more distant the galaxies, the younger and more crowded together they look. Ultimately our view is blocked by the cosmic background radiation, which screens the Big Bang and the cosmic horizon. Cosmologists sometimes refer to this time-distorted view of the observed universe as our "world picture."

But this view is an illusion. Our cosmic horizon has no more physical reality than the familiar horizon observed from a ship at sea. The horizon at sea is an artifact due to the curvature of the Earth, and its location differs for every observer. Likewise, our cosmic horizon is an artifact of observation due to the finite speed of light and the finite age of the universe. Just as every ship at sea appears to be at the center of its own earthly horizon, every galaxy appears to be at the center of its own cosmic horizon.

But as Edward Harrison explains in the third essay of this section, there is another view of the universe that is both more fundamental and easier to grasp than our world picture. Figure 2 schematically represents this view, which cosmologists call the "world map." It depicts the universe as it really is at the present moment of cosmic time — as we would see it if only the speed of light were infinite. All parts of the world map are represented as they exist right now, 13 billion years after the Big Bang. The most remote parts of the map are just as old and evolved as our cosmic neighborhood, with the galaxies spaced just as far apart as in our vicinity.

The world map is uniform, with no center and no horizon. It may well be infinite. All parts of it evolve more or less uniformly together through cosmic time. At any earlier epoch, the world map would show all the galaxies to be younger by the same amount and closer together. We can thus represent cosmic evolution through time as a continuous sequence of world maps, extending all the way back to the Big Bang.

The cosmologist Edward Milne introduced the distinction between the observed world picture and the "real" world map in 1935, but the concept still remains virtually unknown outside the professional community. It provides such a simple and powerful way to understand cosmology and the cosmic horizon that we asked Professor Harrison to contribute an essay on the subject. This essay is the only one in the collection that does not deal with a recent development in astronomy, but its importance justifies making an exception here.

Cosmology raises ultimate questions and some of them may be fundamentally unanswerable. What lies beyond our cosmic horizon? What happened before the Big Bang? Cosmology also stretches both our observational and conceptual limits. A few astronomers question the evidence for the prevailing view that our universe was born in a single explosion from a superdense point. Among the large majority who accept the Big Bang as the origin of our universe, some are exploring theories of multiple self-reproducing universes, in which the Big Bang would have been far from unique.

We conclude this section with a speculative essay in that vein. Lee Smolin describes his hypothesis that black holes produced by massive stars give birth to other universes hidden from our view, and that the Big Bang at the birth of own universe occurred inside a black hole in an older universe. He then asks what would happen if the physical constants that characterize a universe are slightly altered in transmission from a parent universe to its offspring. He concludes that such "mutations" in a sequence of self-reproducing universes would lead to something like natural selection in biology. The process would fine-tune the physical constants so as to maximize the production of black holes and of universes with just the properties of our own. The exquisite complexity of the observed universe would then be the product of many generations of universes, just as the otherwise inexplicable richness of our familiar biological world is a natural result of refinement through long evolution.

To explore the universe, we pose the following questions:

What is the cosmos like on the largest scale we can see?

Michael A. Strauss, Associate Professor in the Department of Astrophysical Sciences at Princeton University, describes how superclusters of galaxies appear arrayed along networks forming "bubbles and voids"— the large-scale structure of our universe.

The expansion of space may be accelerating. Could an unknown force of physics be at work?

Robert P. Kirshner, Professor of Astronomy at Harvard University and an Associate Director for Optical and Infrared Astronomy at the Harvard–Smithsonian Center for Astrophysics, explains how he and his colleagues use exploding stars in distant galaxies to trace the expansion of the universe back in time. The results point to an accelerating universe and perhaps a mysterious force that opposes gravity.

What lies beyond the "cosmic horizon"?

Edward Harrison, Distinguished University Professor Emeritus of Astronomy at the University of Massachusetts, describes how our observed "picture" of the universe is really a composite image of space and time. The picture has a horizon, centered on us. He then introduces the far simpler "map" of the universe as it "really" exists everywhere at the same instant in cosmic time. This view unmasks our cosmic horizon as an illusion.

Are there other universes? Could they be related to ours?

Lee Smolin, Professor of Physics in the Center of Gravitational Physics and Geometry at Penn State University, explains his speculative hypothesis that our universe was born from a black hole in an older universe, and that its physical properties have evolved through countless previous generations of universes.

Mapping the Universe

Michael A. Strauss

The Milky Way Galaxy in which we live is not isolated in the cosmos, but is part of a small group of roughly forty large and small galaxies, which we affectionately call the Local Group. The nearest large galaxy is called the Andromeda Galaxy. It is about 2 million light-years away, which sounds like quite a lot, until we remember that our own galaxy is 100 thousand light-years across. Unlike stars, which are separated from one another by distances incredibly vast compared to their sizes (comparable to one grain of sand every forty miles), our Galaxy is separated from its nearest large neighbor by a distance only twenty times

Michael A. Strauss is an Associate Professor in the Department of Astrophysical Sciences at Princeton University.

Figure 1: The core of the rich galaxy cluster Abell 2218, as seen by the Hubble Space Telescope. This cluster is so massive and compact that it warps the surrounding space, acting like a gravitational lens. This deflects light rays from more distant galaxies, producing the arc-like pattern. The arc segments are actually the images of much more distant background galaxies, but magnified, brightened, and distorted by the lensing effect.

its size. Once you get used to the sizes of galaxies, the universe is not such a terribly large place.

To look beyond the Local Group requires observations with powerful telescopes. With these, we discover that we live in a universe filled with galaxies. Astronomers once assumed that galaxies were distributed more or less randomly in space. But in fact, galaxies, like people, are gregarious. They congregate in groups and clusters. Figure 1 shows a picture taken with the Hubble Space telescope of one of the richest clusters of galaxies known to us. It consists of well over one thousand giant and dwarf galaxies in a region a few million light-years across.

What happens if we look on yet larger size scales? One of the great discoveries of the last two decades is that the distribution of galaxies in space shows structure on scales substantially larger than clusters. The method that allows us to map this distribution is based on the fact that we live in an expanding universe. Light waves traveling across expanding space are stretched to longer wavelengths. The farther a galaxy is from us, the more its light waves are stretched. Measuring the magnitude of this effect allows us to calculate the galaxy's distance. Repeating this for many galaxies yields a map of the distribution of galaxies in three dimensions.

As astronomers mapped the distribution of galaxies in space, they discovered amazing structures: superclusters, walls, filaments, voids, and bubbles. Figure 2 shows the positions of 1,500 galaxies in the Northern Sky measured by John Huchra, Margaret Geller, and

their collaborators. Every dot represents the position of a large galaxy, each containing about 100 billion stars. In this striking map of the galaxy distribution (nicknamed "The Stickman"), we can see coherent filaments (the Stickman's arms) and large empty regions, or bubbles, which seem to be completely devoid of galaxies. Notice the Great Wall, which stretches 500 million light-years across the survey, and the voids which punch all the way through the surveyed volume. Indeed, the largest structures stretch from one end of the map to another. Can we yet conclude that we have found an upper limit to the size of structures in the universe? If we were to map a larger region, would we find yet larger structures?

Astronomical observations plus theoretical modeling tell us that the universe was born roughly 14 billion years ago in the Big Bang, and has been expanding and cooling ever since. Albert Einstein postulated that the universe was initially very smooth, or homogeneous. We now have direct observational evidence that this is the case. We can actually observe the remnant radiation from the initial very dense hot phase in the early universe. If we look with telescopes sensitive to microwaves in every direction in the sky, we see the echoes of the universe's violent beginning: the cosmic background radiation. The universe was once extremely hot, but it cooled as it expanded, until the radiation from the original fireball now appears as a faint microwave glow.

In 1989 NASA launched the Cosmic Background Explorer (COBE) satellite to map this faint radiation with exquisite precision. Figure 3 shows the resulting data. These elliptical maps are projections of the full sphere of the sky onto a flat figure. The upper panel shows the map at normal contrast; this shows the initial smoothness that Einstein predicted. If we increase the contrast to one part in a

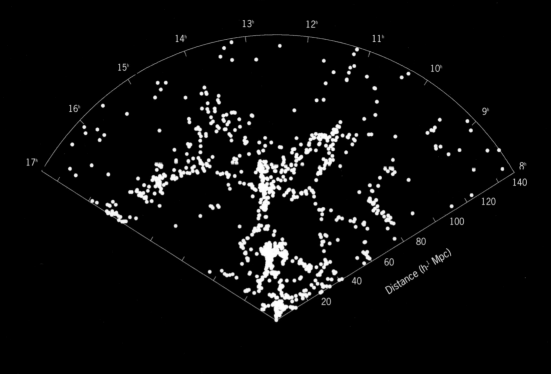

Figure 2: The distribution of roughly 1,500 galaxies in a slice six degrees thick through our universe. Each dot represents a galaxy containing some 100 billion stars. The region shown extends roughly 500 million light-years out from our location at the bottom of the slice. From data of de Lapparent, Geller, and Huchra.

thousand, we get the map in the middle. The pattern we see here is due to the motion of the Milky Way Galaxy through the universe. Once we remove that effect from the data, and increase the contrast by another factor of a hundred, we get the map at the bottom. The band across the middle is due to radio emission from gas clouds in the Milky Way, but the remaining structure is primordial, from a time soon after the Big Bang itself. These fluctuations are very small, only about one part in 100,000.

So our universe started out almost, but not quite perfectly, smooth. And yet today, we know we live in a clumpy universe, which is anything but smooth. We see dramatic structure all around us: people and planets, solar systems, galaxies and clusters of galaxies. On larger scales, the maps of the distribution of galaxies show structures as large as several hundred million light-years across. How did all this structure take shape from such a smooth beginning? In the early universe, the gravitational attraction was slightly higher in those regions of space in which material was slightly clumped, causing more material to be attracted and accumulate there. Thus, these denser regions became larger and more massive, increasing their gravitational pull, attracting yet more matter, and so on. This "snowballing" process, continuing over the 14 billion years to the present, may be sufficient to explain the structures we see today.

In order to explain this process in detail, the theorist creates a computer model of the

Figure 3: The Cosmic Microwave Background Radiation, as observed by the COBE satellite. These elliptical maps are projections of the full sphere of the sky onto a flat figure. The upper panel is the microwave sky at normal contrast, showing no structure. By stretching the contrast to one part in a thousand, we get the map in the middle. The resulting pattern reflects the motion of the Milky Way Galaxy through the universe. Removing that effect from the data, and increasing the contrast by another factor of a hundred, we get the map at the bottom. The band across the middle is due to microwave emission from gas clouds in the Milky Way, but the remaining structure is the distant cosmic background, from a time soon after the Big Bang itself. These fluctuations are very small, only about one part in 100,000.

universe, with various ingredients. The mass of the universe seems to be dominated by dark matter, whose presence can only be detected indirectly by the influence of its gravity on visible matter (see the essay by Christopher Stubbs in Section Three). So we start with a model universe filled with dark matter, and add a small amount of ordinary matter, of which stars, planets, and people are made. We lay down this matter almost, but not quite, perfectly smooth, with just enough clumping to match the structure observed in the COBE maps. The exact form of the clumping we put in depends on the sort of dark matter we are using. (Needless to say, we don't know what the dark matter is made of because we can't directly see it; the physical nature of dark matter is one of the big unanswered questions facing cosmologists today.) Then we let gravity act, and allow our simulated universe to expand and cool for 14 billion years while the clumps grow and structures form. At the end, we look at the structures that have formed, and compare them with what we see in the real universe.

Figure 4 shows slices from these models, at roughly the same scale as the data shown in Figure 2. The left panel shows the results from a model in which the dark matter moves relatively slowly at the time galaxies are forming; this is known generically as Cold Dark Matter. The model and the real universe show qualitatively the same structures: filaments and bubbles, giving us reason to believe that the Cold Dark Matter model is on the right track. The right panel shows the result of a simulation in which the dark matter is made up of ghostly particles called neutrinos—a form of Hot Dark Matter—that move close to the speed of light. This model goes overboard: although we see filaments and bubbles here as well, the model forms enormous clusters, much bigger than anything we observe in the universe today. So

we have learned something by this process: neutrinos cannot be the dominant constituent of the dark matter. It turns out that when one makes these comparisons carefully and quantitatively, even the standard Cold Dark Matter model does not do a perfect job—it does not form structures quite as large as those observed. Moreover, the model does not develop voids that are as empty as what we see in the real universe. Cosmologists are hard at work developing new alternative theories.

How much dark matter is there? This is of course an important ingredient to our models, but it has a deeper significance. We live in an expanding universe. However, every bit of matter in the universe attracts every other bit gravitationally. This means that gravity is acting to pull things together, counteracting the expansion. So if the average density of material in the universe is high enough, its gravity may be strong enough to halt the expansion, reverse it, and eventually lead to collapse. In such a "big crunch," all the material of the universe would eventually collapse down to a single unimaginably dense state. If there is somewhat less matter in the universe, then the expansion will be slowed, but it will not stop, and the universe will last forever, expanding for eternity. If there is a cosmological constant (a hypothetical source of negative pressure that operates over very large scales), then the expansion of the universe may be accelerating, as explained in the essay by Robert P. Kirshner in this section.

Thus we can get a handle on the future fate of the universe by measuring the density of matter within it. If that density turns out to be greater than a certain critical density, then the self-gravity of the universe is strong enough to eventually halt the expansion, and the universe will recollapse. If it is less than the critical density, then the universe will expand forever.

Cold Dark Matter

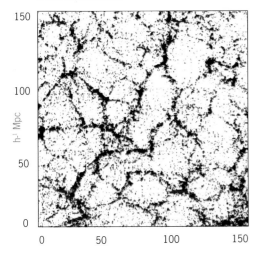

Hot Dark Matter

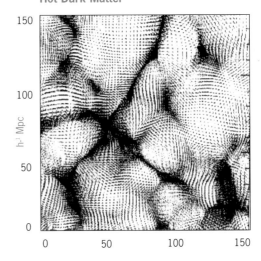

Figure 4: Computer simulations of the galaxy distribution in universes dominated by Cold Dark Matter (left) and Hot Dark Matter (right). The hot dark matter model forms structures larger than those observed in the universe. In the cold dark matter model, structures are not quite as large as those observed, but a closer fit. These are slices through the model, on a scale comparable to the observed galaxy distribution in Figure 2.

How can we "weigh" the universe? We can estimate the mass of each galaxy by measuring the orbital motions of its stars. This, together with the counts of galaxies, gives an estimate of the total mass of the universe. When we do this, we find that the material in galaxies is only about ten percent of that necessary to reverse the expansion, indicating that we live in a universe destined to expand forever.

However, there is an important catch. We know that most of the mass in galaxies is dark matter. Might there be dark matter in the space between the galaxies, which is not even taken into account in this calculation? For example, the great bubbles we've seen in the galaxy maps might be filled with dark matter. Detailed analyses of the motions of galaxies under their mutual gravity is one way to get a handle on this.

A decade ago or so, it seemed that as astronomers observed at larger and larger scales, they found evidence for more and more matter, and the inferred value of the density of the universe seemed to be approaching the critical value. The observations imply that on scales much larger than clusters, there are no further components of dark matter to contribute to the overall density. It seems we live in a universe which has a density roughly one-third that of the critical value needed to halt the expansion. Now, it turns out that a density exactly equal to the critical density is a natural prediction of theories of how the universe formed, while a value close, but not equal, to the critical density, as we seem to be finding, is puzzling. This puzzle is telling us that there is something really basic about how our own universe came to be that we don't yet understand.

What does the future hold? Experience has shown that we always learn new things when we can do surveys in new ways or on larger scales. A large consortium of universities is carrying out a massive survey, called the Sloan Digital Sky Survey, to measure the positions in space of a million galaxies. A million galaxies is a large number in this game. For comparison, the largest such surveys currently in existence include only 20,000 galaxies. The simulation in Figure 5, showing only six percent of what the survey will eventually cover, is an attempt to model what the survey will see, although of course this is only an educated guess. This figure extends to 1.5 billion light-years, which is three times the scale of Figure 2. With a much greater quantity of data in hand over a much greater volume, we will be able to characterize the large-scale structures of the universe with high precision, and thus learn much about both the evolution of the universe, and the formation of the structures it contains.

Figure 5: A computer simulation (assuming that the dark matter is cold) of six percent of the data that the Sloan Digital Sky Survey will obtain in the course of its mapping of one million galaxies.

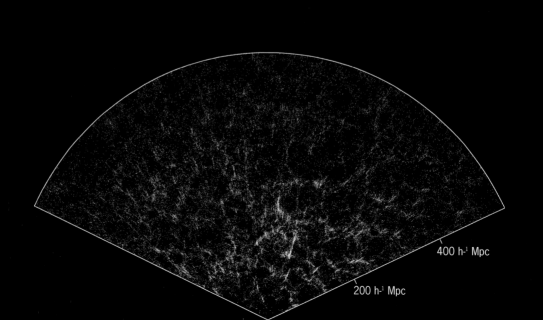

400 h⁻¹ Mpc

200 h⁻¹ Mpc

The cosmic microwave background radiation is the faint
remnant glow of the big bang. This false color image,
covering about 2.5 percent of the sky, shows fluctuations in
the ionized gas that later condensed to make superclusters
of galaxies. BOOMERANG Project.

The Cosmic Microwave
Background Radiation

In 1929, Edwin Hubble showed that the light
from distant galaxies is shifted to longer
wavelengths in proportion to their distances
from the Milky Way. The modern interpretation
is that space itself is expanding, carrying the
galaxies along for the ride. In 1931, Georges
Lemaître imagined running such an expansion
backwards in time (see his Profile in this
section). At some remote point in the past, he
reasoned, everything in the universe would have
been packed together at enormous density.
Lemaître suggested that all the matter and
energy in the observable universe originated in
an explosion of space, now called the Big Bang,
which launched the expansion that continues to
this day.

In 1948, Hermann Bondi, Thomas Gold, and
Fred Hoyle published an alternative

cosmological theory, which accounted for the
observed expansion without invoking a
beginning in time. They proposed that matter is
continually created, to form new galaxies, so
that the expanding universe maintains the same
average density and appearance through infinite
time. In this "steady state" theory, matter is
created continuously. In the Big Bang theory,
all the matter in the universe is created at once,
at a definite point in the past.

In the same year, the physicists George
Gamow, Ralph Alpher, and Robert Herman
developed a detailed theoretical picture of the
Big Bang. They realized that the universe
immediately after the explosion would have
been not only extremely dense but also
extremely hot. At such high temperatures
most of the contents of the universe would be

in the form of intense light (radiation) rather than in the form of matter. This early period is now called the radiation era.

As the universe expanded, the total amount of light and matter had to fill a continually increasing volume of space, so the density of each had to decrease. But the expansion of space also stretched out the waves of the light traveling through it. And the longer the wavelength of light, the lower its energy. So the expansion of space caused the energy density of light to decrease even faster than the density of matter. Consequently, most of the energy of the universe was soon in the form of matter instead of radiation, and today we live in a matter-dominated universe.

The three scientists recognized that the radiant energy of the Big Bang must still exist in the universe today, although greatly reduced in intensity by the expansion of space. Alpher and Herman went on to calculate the present temperature corresponding to this energy. The answer they got was 5 K, which means 5 degrees above absolute zero on the Kelvin scale. (At absolute zero, the lowest possible temperature, molecular motion and thermal radiation come to a complete stop.) Radiant energy at a temperature of 5 K is mostly in the frequency band of microwaves.

Alpher and Herman in effect predicted that the universe today should be awash in a faint but uniform bath of microwave energy coming from every direction—the remnant glow from the Big Bang. But they made no attempt to search for it. As theoretical physicists, not observational astronomers, they perhaps assumed that the technology required for such an observation did not yet exist. Furthermore, radio astronomy was in its infancy in those days, and the handful of radio astronomers who might have known how to use the available technology to search for the microwave background radiation were

unaware of the published theoretical prediction. So for several years the debate between the steady state and Big Bang theories continued, in the absence of any strong observational evidence in favor of one over the other.

In 1964, Arno A. Penzias and Robert W. Wilson at the Bell Telephone Laboratories in New Jersey began investigating the microwave radio emissions from the Milky Way and other natural sources. They had a very sensitive detector connected to a large horn-shaped antenna, previously used for satellite communication. When the two scientists tuned their equipment to the microwave portion of the spectrum, they discovered an annoying background static that wouldn't go away. No matter where they pointed the antenna, or when, the microwave static was the same. They spent months running down every possible cause for the static, including pigeon droppings inside the antenna, but they couldn't find a source or a solution.

At about the same time, Princeton physicist Robert H. Dicke had come to his own conclusion that residual radiation from the Big Bang must still be present in the universe. He did not know about the previously published work by Gamow, Alpher, and Herman. So Dicke independently calculated that the lingering radiation should have a temperature of about 10 K. He realized that it should be observable in the microwave portion of the spectrum. His research team was in the process of building an antenna to search for it when he learned that Penzias and Wilson had discovered a persistent microwave background noise. Dicke turned to his colleagues and said simply, "They've got it."

Penzias and Wilson had stumbled on the first observational evidence to support the Big Bang theory of the origin of the universe. For this discovery they shared the Nobel Prize for Physics in 1978. Subsequent observations of the microwave background at different wavelengths have refined the value of the radiation temperature of the universe to 2.73 K. This is about half the value calculated by Alpher and Herman in 1948, but their result is widely regarded as a successful prediction in view of the approximations required by the calculation. The discovery of the cosmic microwave background radiation led most astronomers to accept the Big Bang theory.

The few who dissent from Big Bang theory include, among others, the authors of the original steady state theory. They suggest that ordinary starlight, not the Big Bang, produced the microwave background radiation. If this were true, there would have to be a mechanism, as yet unverified, to convert the visible starlight into the observed microwave spectrum. The dissenters continue to investigate such possibilities.

Arno Penzias (right) and Robert Wilson by the horn shaped antenna they used to discover the cosmic microwave background radiation.

Cosmic Expansion and Acceleration

Robert P. Kirshner

Our solar system resides in one of a hundred billion galaxies in a universe that has been expanding for about 14 billion years since the Big Bang. We are now detecting clues that the expansion may be accelerating, driven by some strange properties of space itself. If current interpretations are correct, most of the universe is not in the form of luminous matter or even of dark matter detected only by its gravitational effect, but in a strange form of energy associated with the vacuum of space. What is the evidence for such a bizarre situation? Let me sketch the facts as we know them.

Robert P. Kirshner is Professor of Astronomy at Harvard University and Associate Director for Optical and Infrared Astronomy at the Harvard-Smithsonian Center for Astrophysics.

Early in the twentieth century, astronomers had established that the Sun is embedded in a big system of stars and gas we call the Milky Way. But the faint cloud-like spiral nebulae remained enigmatic. What was their relation to the Milky Way? Was the Milky Way the entire universe, or were the spirals perhaps systems as grand as the Milky Way but very far away? In the 1920s, Edwin Hubble computed distances to the spiral nebulae by identifying and measuring the apparent brightness of luminous supergiant stars within them. Knowing the true luminosity of such stars, Hubble was able to deduce that the spirals must be millions of light-years away. This placed them well outside our own Milky Way Galaxy and led to a picture of our Galaxy as one among many, more-or-less equivalent, systems.

Hubble also measured a shift to longer, redder, wavelengths of features in the spectra of these galaxies. He found that the more distant a galaxy, the larger the shift of its light toward the red end of the spectrum. We now know that space itself is expanding, stretching out the waves of light as they travel across it. The expanding space carries the galaxies along with it. The more distant a galaxy, the more the light we receive from it is redshifted. This is Hubble's Law: distant galaxies are moving away from us and the distance is proportional to the redshift. A plot of redshift (or velocity) versus distance is called a Hubble diagram.

What is the meaning of Hubble's Law? The fact that all the galaxies are receding from us might suggest that we are at the center of the universe. But the Earth is not at the center of the solar system, the Sun is not at the center of the Milky Way, and our Galaxy is not at the center of the universe. It only looks that way, but it looks that way from every galaxy.

If the whole of space is stretching out in every direction, if the universe is expanding, then each observer will see the others receding according to Hubble's Law, with the recession velocity

Cosmic Redshift

The expansion of space stretches out the wavelengths of light, and accounts for the cosmic redshift. At Time 1, a distant galaxy emits light with a short wavelength. The cube shows a typical volume of space at that time, which contains the light wave. At Time 2, the universe has doubled in size (shown by the expanded cube), and the same light wave has doubled its wavelength. At Time 3, the universe has tripled in size and the wavelength of the light received by an observer is now three times longer than when emitted by the galaxy.

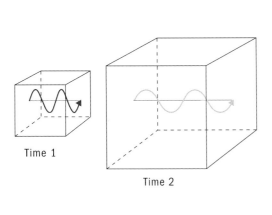

Time 1

Time 2

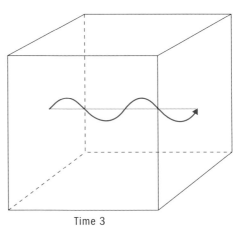

Time 3

A   B   C

Time

Slices of the universe through time.
Three models for the expanding universe. Time increases from bottom to top. Sequence A follows a two-dimensional section of a universe that expands at a constant rate. Each circle contains the same group of galaxies as space expands. Sequence B shows four successive stages in a universe where the expansion is slowing down due to gravity. In such a universe, the expansion rate is initially rapid and then slows down. The distance between galaxies initially increases rapidly, and then more slowly. Sequence C shows four stages in a universe where the expansion is accelerating, perhaps due to a cosmological constant that speeds up the expansion of space. The expansion rate is initially slow and then increases. Such a universe might expand forever.

proportional to distance. Imagine a balloon covered with ants. If you blow up the balloon, each ant will see its neighbor apparently moving away as the space between them stretches, and the stretching will add up so that more distant ants are moving away more rapidly. But no ant is at the center of the expansion. Every one would see the same recession law.

While the expansion of the universe was found by careful observations of stellar brightnesses and galactic spectra, it was theoretically foreshadowed a decade earlier. As Albert Einstein worked out his theory of gravity, called General Relativity, he found solutions to his equations that corresponded to an expanding universe or to one that contracted, but there were no solutions that gave a static universe. Also, according to the theory, if the average density of the universe were less than a particular "critical" value, the universe would expand forever, and if the density were greater than that value, even an expanding universe would eventually turn around and start contracting.

Einstein asked the astronomers what the evidence was concerning an expanding or contracting universe. In 1920, "universe" meant "Milky Way Galaxy," since Hubble had not yet established the distances to the spiral nebulae. So the reply was, "The universe is neither expanding nor contracting—it is static." Einstein took this "fact" seriously. He added a "Cosmological Constant" to his equations for the express purpose of making the universe static. The Cosmological Constant acts like a repulsion that, in effect, cancels out the gravitational tendency for the universe to contract. If he had resisted this temptation, Einstein would have predicted the expanding (or contracting) universe. He later referred to the Cosmological Constant as "my greatest blunder." But, as we will see, we may yet need it to account for recent data on the expansion of the universe.

The luminous supergiant stars that Hubble observed in nearby galaxies with the 100-inch telescope at Mount Wilson can now be measured in much more distant galaxies with the Hubble Space Telescope. The HST is no larger, but it does have a superior site, above the Earth's atmosphere. Even with the HST, however, we cannot yet determine the distances to galaxies more than 100 million light-years away—about one percent the size of the observable universe—by observing the brightest stars they contain. To survey the cosmos over larger distances, and to probe its expansion at earlier epochs, we need a brighter source of known luminosity, a brighter so-called "standard candle."

The supergiant stars that Hubble could see in other galaxies are about 10,000 times more luminous than the Sun. Although they are bright stars, they are dim bulbs compared to the light emitted by a dying star in a supernova explosion. Supernovae are of two types: Type I are due to the thermonuclear detonations of compact white dwarf stars, and Type II result from the gravitational collapse of massive stars. The Type I supernovae have more uniform properties that make them suitable as distance measuring tools for cosmology. A typical Type I supernova is about four billion times more luminous than the Sun—comparable to the light output of an entire galaxy. Figure 1 shows supernova 1994D in the outskirts of a disk galaxy called NGC 4526, as recorded by the HST.

Since supernovae are so bright, we can see them far out into the expanding universe, where the recession velocities are large, and measure the ratio of their velocity to distance. The rate of expansion measured in this way can tell us how long the universe has been expanding.

It is easier to visualize using the analogy of a long footrace, like a city marathon. After some initial jockeying for position, the race stretches out along the route. If you are somewhere in the middle of the pack, there will be faster people up the road ahead and slower people behind you. As time passes, the faster runners get farther ahead of you and the slower runners (if there are any) get farther behind. You can compute how long the race has been underway if you know the relation between velocity and distance. If you see somebody one half-mile up the road, and you know they are running one mile per hour faster than you, then you could figure that the race had been underway for one half-hour.

The same is true for the expanding universe: there is an "expansion time" that tells how long it would take a galaxy to get to its present distance from us if space constantly expanded at its present rate. This time is the same for all galaxies because the whole universe started at the same time and the velocity is proportional to the distance: a galaxy at twice the distance has twice the speed, so the time, which is distance divided by speed, is the same. Our observations of the distances and speeds of galaxies indicate that the expansion has been underway for about 14 billion years. Should we take this seriously as the age of the universe?

One way to find out is to compare the expansion time with the ages of objects in the universe. If it really marks the time since the beginning, then we should not find anything older! We can calculate the age of the oldest stars in our Galaxy by observing globular star clusters containing low mass stars that are just swelling up to become red giants. From their luminosity, we can determine the mass of those stars. We know how long it takes a star of a given mass to use up its hydrogen fuel and begin the change to a red giant. Our best estimates for the ages of globular clusters range from 12 to 15 billion years, which is in reasonable agreement with the expansion age of the universe.

However, observations of a marathon reveal an omission in the way we computed the expansion age. Runners get tired and slow down. So, if you judged the duration of the race from the pace of the last five miles, you would over-estimate the duration of the race. In the same way, if galaxies gradually slow down due to the mutual attraction of their gravity, the simple estimate of 14 billion years, which assumes a constant expansion rate, would be misleading. For example, if the universe had exactly the critical density, which separates models that expand forever from those that eventually collapse, the correction for deceleration would make the time since the Big Bang only 9 billion years. This is uncomfortably short—less than the age of the oldest stars. But the stars can't be older than the universe, any more than you can be older than your mother!

There are two ways to assess the effects of cosmic deceleration due to gravitation. One is to measure the average density of the universe. While this is tricky (since most of the mass in the universe is dark, not luminous), the best estimates show that the density of all the mass adds up to only about thirty percent of the critical density needed to halt the present expansion. But these methods measure only the density of matter clumped into galaxies, galaxy clusters, and the filaments that surround cosmic voids (see the essay by Michael A. Strauss in this section). They do not measure the effects of anything smoothly distributed. Another way to measure the density is to observe the deceleration directly. In a marathon, you could compare a runner's time over the first five miles with that over the last five miles to see if the runner was slowing down. In the cosmic expansion, we can measure the relation between redshift and distance for nearby galaxies and compare it to the redshift-distance relation far away (and long ago, since the light from an object halfway across the universe takes billions of years to reach us). The difference might tell us the deceleration.

Type I supernovae are our best tool for this job. They are bright enough to see halfway across the universe. Unfortunately, they are rare. We must search about 5,000 galaxies to catch such a supernova within a week of its maximum brightness. Fortunately, the technology for doing this has matured in the past few years; giant electronic cameras on large telescopes can image an area as big as the full moon down to very faint limits of brightness. By imaging the same areas of sky a few weeks apart, the telltale signs of distant supernova eruptions can be picked out by subtracting one image from another in a computer. By measuring the brightness of a distant supernova, we can compute its distance. By obtaining a spectrum, we learn the redshift and extend the Hubble diagram to look for the signature of cosmic deceleration.

If the expansion of the universe has been slowing down while the light from a distant supernova was on its way to our telescope, then the light from the supernova will have

Evidence of expansion and acceleration.
A plot of distance versus redshift of galaxies provides
evidence for expansion and acceleration of the universe.
Each point represents a different galaxy, with its distance
based on the apparent brightness of a supernova. The
nearer galaxies, in the lower left part of the plot, fall near
a straight line, and the distance of each galaxy is directly
proportional to the observed redshift. This is Hubble's Law,
the evidence for an expanding universe. For redshifts
greater than about 0.1, the line curves slightly upwards,
due to the effects of general relativity. This line represents
a theoretical model of the universe that has the minimum
observed mass and no cosmological constant to cause
acceleration. The most distant galaxies fall distinctly above
this line, so they appear to be farther away and fainter
than predicted. This suggests that the expansion rate
of the universe increased while the light from those
galaxies was on its way to us, so the light had to cover
more distance than if the universe were expanding at a
constant rate.

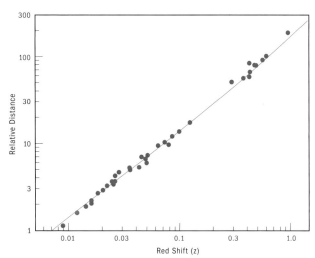

covered a little less distance than if the universe
has been coasting with a constant rate of
expansion. This is something like throwing a
snowball at a departing schoolbus. If the bus is
slowing down, the snowball will not have to
travel as far to make a satisfying "splat."

Since a decelerating galaxy is closer, its light
would appear a little brighter than a galaxy
with the same redshift in a coasting universe.
In fact, distant supernovae should look about
twenty-five percent brighter in a decelerating
universe with critical density (often assumed as
a standard model) than in a universe expanding
at a constant rate.

In 1998, two teams reported results on distant
supernovae—the Supernova Cosmology Project
led by Saul Perlmutter and the High-Z Supernova
Team led by Brian Schmidt, of which I am a
member. Both had adequate samples of
supernovae in galaxies to see the expected
twenty-five percent increase in brightness.
Both groups got a big surprise. The distant

supernovae were not brighter than you'd
expect in a constant-expansion universe. They
were fainter, and by a significant amount.
If this dimming is due to the way the universe
is expanding, it implies the opposite of
deceleration, that is, acceleration. If the
expansion is accelerating, the supernova
observations are the first experimental
indication of a very important physical effect.

Of course, there could be another cause for the
observations coming out as they did. Perhaps
the more distant supernovae are dimmer
than nearby ones because they are (on average)
from younger stars, or they have a different
chemical composition, or their light suffers
more absorption by some intergalactic dust.
Both teams looked into these possible
problems and concluded that they probably
do not dominate the measurements.

If we take the observations seriously, they
point toward an extraordinary conclusion. Since
ordinary gravity is only attractive, and only

produces cosmic deceleration, you need to add something to the equations of General Relativity to account for the observations. Einstein's discarded Cosmological Constant has just the right properties. It makes the universe expand more rapidly over time, just as the supernova observations indicate. If this is right, then we have just made a conceptual transition from a universe dominated by (dark) matter into one controlled by the properties of the vacuum energy. The simplest prediction is that the expansion of the universe we see today will continue, and continue to accelerate forever.

The Cosmological Constant acts like an energy associated with the vacuum of empty space. This is not as wacky as it sounds, since particle physics is based on quantum theory, in which the vacuum is a lively place, full of "virtual" particles being created and destroyed. But no laboratory experiment, even on the grandest scale, has given us any clue about the value of the Cosmological Constant. Only astronomical observations can reveal this part of the physical world—and it may be the largest part!

The universe has about thirty percent of the critical density in matter, as determined from the gravitational effects of galaxy clusters, and the best estimate of the Cosmological Constant (based on the acceleration of the expansion) suggests that it corresponds to about seventy percent of the critical density. Though these numbers are only approximate, they indicate that the energy associated with the vacuum is larger than that associated with all forms of matter, visible or invisible. These two numbers also add up to the critical density, which means that the universe would have the simplest geometry permitted by General Relativity, the Euclidean geometry of flat space.

This is significant, because a fruitful theory of the Big Bang, called "inflationary cosmology,"

predicts that space should be flat. So perhaps the pieces are beginning to fit together: We may have a universe that began with inflation, then settled down to a decelerating era dominated by gravity from mass-energy, and is now entering an era when the vacuum energy governs the expansion.

Lest we get too proud of this creation, we should remember that we don't know what the dark matter is made of, and we don't know whether the Cosmological Constant is the true source of the acceleration observed in the Hubble diagram for distant supernovae. And it is always possible that something less exciting, like some unfamiliar kind of intergalactic dust, is misleading our interpretation of the data. We need to test this picture—by observing supernovae at still higher redshifts, and by combining the supernova data with observations of the cosmic microwave background radiation and other clues from the time when our universe was young. New telescopes, such as Gemini and Magellan, and new satellites, like the Microwave Anisotropy Probe (MAP) and the Next Generation Space Telescope (see the essay by Alan Dressler in Section Six), will help us test these ideas.

Though we are feeble little beings, I think it is permissible for us to take some pride in our collective achievement. The origin, age, and fate of the universe have been matters for legend and myth for thousands of years. Now we are beginning to understand enough to begin to answer these questions through experimental science. Nobody should assume the present answers are the definitive ones: our knowledge comes down from the mountain on magnetic tapes, not stone tablets. But the questions "Where are we?," "What time is it?," and "Where are we are going?" are worth answering, and we are beginning to make some progress.

Fritz Zwicky's Extraordinary Vision: Profile

Fritz Zwicky (1898–1974), Swiss-American astronomer who first conceived of supernovas, neutron stars, dark matter, and gravitational lensing by galaxies.

If ever a competition were held for the most unrecognized genius of twentieth century astronomy, the winner surely would be Fritz Zwicky (1898–1974). A bold and visionary scientist, Zwicky was far ahead of his time in conceiving of supernovas, neutron stars, dark matter, and gravitational lenses. His innovative work in any one of these areas would have brought fame and honors to a scientist with a more conventional personality. But Zwicky was anything but conventional. In addition to his brilliant insights that turned out to be right, he also entertained notions that were merely eccentric. To his senior colleagues he could be arrogant and abrasive. He referred contemptuously to "the useless trash in the bulging astronomical journals." He once said, "Astronomers are spherical bastards. No matter how you look at them they are just bastards." His colleagues did not appreciate this aggressive attitude and, mainly for that reason, despite Zwicky's major contributions to astronomy, he remains virtually unknown to the public.

Zwicky was born in Bulgaria of Swiss parents. He earned his Ph.D. in physics from the Federal Institute of Technology in Zurich, and then spent most of his life in the United States. He joined the California Institute of Technology in 1925 to work on the physics of crystal structure. But he was soon caught up in the excitement of astronomical research at Caltech. He became fascinated with cosmic rays—the high energy subatomic particles that shoot through space at nearly the speed of light. No one could suggest a plausible candidate for the source of the mysterious particles. Then Zwicky made an astonishing conceptual leap. He decided that cosmic rays are produced in catastrophic explosions of massive stars. No one had previously imagined such a phenomenon. In a 1931 lecture course at Caltech, Zwicky introduced the term "super-nova" to distinguish the explosion of an entire star from the more common and much less powerful nova, which involved violent and repeated outbursts on the surface of an unstable star.

Zwicky teamed up with the German-American astronomer Walter Baade to work on the supernova idea. Baade knew of several historical accounts of "new stars" that had appeared as bright naked eye objects for several months before fading from view. The Danish astronomer Tycho Brahe, for example, had made careful observations of one in 1572. Zwicky and Baade thought that such events must be supernova explosions in our own Galaxy. At a scientific conference in 1933, they advanced three bold new ideas: (1) massive stars end their lives in stupendous explosions which blow them apart, (2) such explosions produce cosmic rays, and (3) they leave behind a collapsed star made of densely-packed neutrons.

Zwicky reasoned that the violent collapse and explosion of a massive star would leave a dense ball of neutrons, formed by the crushing together of protons and electrons. Such an object, which he called a "neutron star," would be only several kilometers across but as dense as an atomic nucleus. This bizarre idea was met with great skepticism. Neutrons had only been discovered the year before. The notion that an entire star could be made of such an exotic form of matter was startling, to say the least.

Zwicky made a persuasive case that supernovas actually occur and ought to be observable in other galaxies. Around 1935, he convinced George Ellery Hale, the Director of Mount Wilson Observatory, to build him an 18-inch Schmidt telescope, which had an unusually wide field of view, ideal for photographing many galaxies at once. In three years, Zwicky used it to discover twelve supernovas. He then persuaded Hale to build the 48-inch Schmidt telescope at Mt. Palomar. Its primary purpose was to photograph the entire northern sky, and the resulting Palomar Observatory Sky Survey became a

major cornerstone of astronomy for the next fifty years. But Zwicky also used the 48-inch telescope for "supernova patrols." He eventually discovered 122 supernovas, still a record for any one observer.

Astronomers readily accepted supernovas, but remained doubtful about neutron stars. Zwicky persisted and was ultimately vindicated. In the late 1930s, theoreticians showed that neutron stars were compatible with nuclear physics. Then, in 1967, radio astronomers discovered the first pulsars, and the next year Thomas Gold of Cornell University showed that such objects could only be rapidly spinning neutron stars produced in supernovas.

Zwicky's independent mind also emboldened him to challenge the general assumption that the mass of the universe consists mostly of stars. In 1933, while investigating the great Coma cluster of galaxies, he stumbled upon a major discrepancy between theory and observation. The average speed of galaxies within a cluster depends on the total mass of the cluster, since each galaxy is attracted by the gravity of all the others. From the observed speeds of galaxies moving within the Coma cluster, Zwicky calculated its total mass. He then added up all the light from the galaxies in the cluster and used it to calculate the mass in the form of luminous stars. To his surprise, the mass of the cluster based on the speed of its galaxies was about ten times more than the mass of the cluster based on its total light output. He concluded that the Coma cluster must contain an enormous quantity of unseen matter, with enough gravity to keep the rapidly moving galaxies from flying apart. Zwicky in effect discovered that most of the mass in the universe is invisible. He called it "dark matter." (See the profile on Vera Rubin in Section Three.)

In 1937 Zwicky thought of another way to investigate dark matter. If by chance a massive galaxy lies along our line of sight to a more distant galaxy, it could act as a "gravitational lens," warping the surrounding space to magnify, distort, and even multiply the image of the background galaxy. This was a direct application of Einstein's Theory of General Relativity. The bending of starlight by the gravity of the Sun had already been demonstrated in 1919. Zwicky predicted that massive galaxies would similarly distort the light rays from background objects and that the distortion could be used to "weigh" the lensing galaxies. Most astronomers did not take this idea seriously. But in 1979, five years after Zwicky died, the first of many gravitational lenses was discovered, and a cottage industry has since emerged to find and study them. The lensing effect is now used to measure the cosmological parameters of the universe, and to reveal distant objects otherwise too faint to see. (See the essay by Michael A. Strauss in this section.)

Zwicky also predicted the existence of low mass galaxies and then discovered the first such "dwarf" galaxies with the 100-inch telescope at Mt. Wilson. He anticipated the discovery of quasars by predicting that compact blue galaxies with high luminosity might be mistaken for nearby stars. He was one of the first astronomers to emphasize that the distribution of clusters in the universe is far from uniform on the largest scale.

A diligent worker, Zwicky published hundreds of papers in a wide range of topics. He produced a six-volume catalog of some 30,000 galaxies based on the Palomar Observatory Sky Survey, which remains a standard reference on galaxy clusters. He also published a catalog of bright compact galaxies, which proved invaluable in leading astronomers to find so-called active galaxies. In the book's introduction, however, Zwicky included an intemperate rant describing other astronomers by name or allusion as "fawners" and "thieves" who stole his ideas and hid their own errors.

What is one to make of such an irascible character? Some of Zwicky's contemporaries regarded him as an irritating buffoon. Others had a more balanced assessment. Jesse Greenstein, former chairman of the Caltech Astronomy Department, said "I disliked him as a human being. He was vain and very self-centered. Zwicky had an enormous facility to produce radical new ideas, some of which proved to be correct, but a lot of us wish he had not been so rough in the process." Greenstein recalled the feud between Zwicky and Walter Baade that blew up during World War II. "Zwicky called Baade a Nazi, which he wasn't, and Baade said he was afraid that Zwicky would kill him. They became a dangerous pair to put in the same room." But Greenstein acknowledged that Zwicky "also had a humanitarian side. He stockpiled astronomy books in the basement of our department after the war. Without my knowledge… he bought these books using department funds and shipped them off to the war-torn astronomy libraries in Europe. This was quite gracious, but it cost us a lot of money, which we didn't have at the time."

Zwicky's combativeness was apparently reserved for his peers. He was friendly toward students and administrative staff, whom he didn't regard as competitors. Younger astronomers today tend to regard him as something of a hero, an eccentric genius who advanced astrophysics by proposing truly brilliant and outrageous new ideas without being worried in the least about the consequences. Such an ability is extremely rare, and we should not necessarily be surprised to find it coupled with so peculiar a character.

Two Views of the Cosmos

Edward Harrison

" How distant some of the nocturnal suns!
So distant, says the sage, 'twere not absurd
To doubt that beams set out at Nature's birth
Had yet arrived at this so foreign world,
Thought nothing half as rapid as their flight. "

—Edward Young, *Night Thoughts on Life* (1741)

Edward Harrison is Distinguished University Professor Emeritus of Astronomy at the University of Massachusetts.

Allegory of the finite celestial sphere and the infinite universe beyond. Despite its Medieval style, this engraving first appeared in 1888 in Camille Flammarion's *L'atmosphere: Meteorologie populaire*.

Pure reason alone assured René Descartes, renowned French philosopher and scientist in the seventeenth century, that light travels at infinite speed. How confusing our perceptions of the real world would be, he argued, if the light from different distances originated at different times! An object at the moment of observation might not be what it seemed, and might have changed or moved or both. To Descartes and many Cartesians who followed, the notion that when we look out in space we also look back in time seemed too preposterous to be taken seriously. The deep-rooted belief that light travels instantaneously (the visual-ray theory) originated in the ancient world and to this day we (astronomers also in our off-guarded moments) unconsciously assume that we see the heavens as they are at the moment of observation.

The Danish astronomer Ole Roemer discovered in the late seventeenth century that when the Earth recedes from Jupiter, the orbital period of Jupiter's moon Io appears to increase, and when the Earth approaches Jupiter the orbital period of Io decreases (see the Profile on Roemer in this section). He was the first to discover the Doppler effect: when we move toward a source of sound or light, the observed frequency increases, and when we move away, it decreases. In this case, the frequency is not of sound or light, but of Io's orbital revolution. Roemer's results, updated by Edmund Halley, showed that the speed of light was about 300,000 kilometers per second. Most Cartesians remained unconvinced. Their opposition collapsed in the early eighteenth century, when James Bradley found that stars appear to move backward and forward as the Earth orbits the Sun, and their small angular displacement (called aberration) confirmed incontestably that light has finite speed. In everyday life, the finite speed of light causes no perceptual confusion. But when distances are very large, as in astronomy and cosmology, Descartes' fears of perceptual confusion are fully realized.

Astronomy in the eighteenth and nineteenth centuries advanced by leaps and bounds. Yet in the same period, cosmology, the science of the universe as a whole, made almost no progress, mainly because of inhibiting religious beliefs. Concepts such as the horizon of the visible universe stayed unexplored, and problems such as why the sky at night is dark in an unbounded universe populated uniformly with stars (a.k.a. Olbers' paradox; see the Case Study on Olbers' paradox in this section) stayed unsolved. Astronomers built larger telescopes, peered deeper into space, and published hundreds of popular works on the vastness and glory of the heavens. But few mentioned that when we look out over large distances in space we also look back through vast periods of time, and none to my knowledge discussed the age of the universe.

Most astronomers, particularly in the English-speaking world, were clergymen and respected members of society. Naturally, they had no wish to take issue with contemporary religious beliefs concerning the age of the universe. According to biblical chronology the universe was created less than 10,000 years ago. Yet telescopes revealed a universe extending millions of light-years in space and therefore existing for million of years. The old visual-ray theory of instantaneous vision, used uncritically, conveniently reconciled astronomical observations and religious precepts. Credit goes to the geologists and paleontologists and later the physicists for showing that the age of the Earth, and therefore of the universe, must be measured not in thousands but in millions and perhaps billions of years.

Remarks on the travel-time of light before the twentieth century are rare gems in the history of astronomy. In 1802, Sir William Herschel wrote in a letter that light rays from a remote nebula must have been "almost two millions of years on their way, and consequently, so many years ago, this object must already have had an existence in the sidereal heavens." In 1901, Lord Kelvin showed that the extent of the visible universe (the region we can see) is determined by the age of the universe, and solved Olbers' paradox by showing that the stars in the visible universe are too few to cover the sky, and therefore they cannot create an inferno of starlight, as argued in the paradox.

Unfortunately for observational cosmology, Descartes was wrong. We must always think of two views of the universe: the universe as it is at this moment everywhere in space, and the universe as we see it stretching away in space and back in time. The English astrophysicist and cosmologist Edward A. Milne in 1935 referred to the first view as the "world map" and the second view as the "world picture." The world map is what we would see if the speed of light were infinite (the Cartesian view), with all parts of the universe seen as they really are at the present instant. The world picture is what we actually see because of the finite speed of light (the Roemer view), with distant objects appearing as they were in the past. Milne proposed in 1933 the celebrated "cosmological principle": all places are more or less alike in the world map.

To make things clear, consider an imaginary gadabout explorer who travels around the universe at infinite speed. The explorer sees the world map, whereas the stay-at-home observer sees the world picture.

The explorer finds that things in the world map are everywhere much the same. Distances are well-defined, even when space is curved, and

are of the tape-measure kind (obtained in principle by stretching a tape measure between points). The explorer rushes around in the world map, setting all clocks to the same time, and periodic tours thereafter reveal that the clocks are running in synchronism. The explorer has thus established a universal standard time (known as cosmic time) that is used in cosmology theory. The world map is the universe at an instant in cosmic time.

The universe expands; actually, the space between galaxies expands and the galaxies—floating in expanding space but not themselves expanding—recede from one another. (Note: the recession velocity is the increase in tape-measure distance in a unit of cosmic time.) From the cosmological principle we find, by logical argument, that the speed at which distant galaxies recede from each other is strictly proportional to distance in the world map. This very important result (hinted at in Hubble's observations) holds true for all tape-measure distances in the world map, no matter how large.

A light ray traveling in expanding space has its wavelengths steadily stretched and its spectrum steadily shifted toward the red. This is the exquisite explanation of cosmological redshifts. The redshift of light from a distant galaxy measures the amount the universe has expanded since the light was emitted. A redshift of 1, for example, means that the universe has increased in size by one hundred percent, and all tape-measure distances have doubled since the time the light was emitted.

The recession velocity between any two distant bodies in the world map increases in proportion to their separation, and becomes equal to the velocity of light at a distance of roughly 15 billion light-years. Light rays at this distance in the world map, moving toward us through receding space, actually stand still! The spherical volume of space around us (or around any observer) with a radius equal to this critical distance is called the "Hubble sphere." Galaxies within our Hubble sphere recede slower than light; their light rays travel toward us across expanding space that is itself receding slower than light. All galaxies outside our Hubble sphere recede faster than light; their light rays, traveling in space that recedes faster than light, actually recede from us. (Relativity theory permits nothing to move through space faster than light, but allows space itself to expand faster than light.) As Arthur Eddington wrote in 1933 in *The Expanding Universe*, in this case "light is like a runner on an expanding track with the winning post receding faster than he can run."

In some cosmological models, the expansion is slowing down and the Hubble sphere is expanding faster than the universe. Galaxies now outside the Hubble sphere will be overtaken in the future and lie inside, and their light rays will then be free to approach and reach the observer.

We cannot observe the world map in which things are nicely uniform and distances are neatly defined. We see instead the world picture in which things are nonuniform and distances are anybody's definition. We observe a universe in which galaxies appear to change with distance because of their evolution in time. At very small redshifts, the world map and the world picture tend to be identical.

To make sense of the universe, we must project the observed world picture onto the theoretical world map. This means that the familiar language of the world picture must be translated into the unfamiliar language of the world map. The distance "then" to a galaxy in the world picture, at the time the light now observed was emitted, is not the same as its

distance "now" in the world map. Distances in the world map are defined with geometrical precision. Distances in the world picture are defined operationally, depending on the type of observation (parallax, luminosity, redshift, and so forth), and must be translated into world map distances.

The Doppler interpretation of cosmological redshifts, popular in the world picture, implies that distant galaxies recede through static space. It must be discarded and replaced in the world map by the expansion-redshift interpretation, which follows directly from the expansion of space itself. The observer's Hubble law, according to which redshift increases with distance (true only for small redshifts), must be translated into the theorist's velocity-distance law, in which recession velocities are strictly proportional to tape-measure distances. Having done all this, and even more, we may attempt to determine how fast and in what way the universe expands.

In many cosmological models, space is bounded in the world picture because light has traveled only a finite distance since the beginning of the universe. But in the world map all places are much alike and space is unbounded. Thus, the "observable universe" is only a small part of the whole universe. We are at the center of an observable region bounded by a horizon beyond which we cannot see. A ship far from land is at the center of a "visible sea" stretching away to a horizon. Roughly, if the universe has a beginning in cosmic time (for example, if it originated as a Big Bang), it has a cosmic horizon. We do not see galaxies beyond the horizon because their light has not yet reached us. In the language of the world map, the horizon sweeps outward past the galaxies at the speed of light, and in the course of time we see more of the universe.

It is now more than seventy years since Edwin Hubble's pioneer work on the expansion of the universe, and we still are not exactly sure how fast the universe expands. The rate of expansion claimed by observers has changed tenfold since Hubble's time and still jumps up and down. Why, we must ask, is observational cosmology so terribly difficult? Distances, of course, are difficult to measure, but is this the only reason?

Galaxies and their clusters move randomly and their motions in local space dominate over the cosmic flow of space itself. This region of local motions is the "sub-Hubble sphere." Inside the sub-Hubble sphere, the peculiar motions dominate; outside, cosmic recession dominates. At what redshift do we quit the sub-Hubble sphere and enter the true cosmic flow? We still do not know.

I have a heretical thought (every cosmologist is allowed one). Perhaps the sub-Hubble sphere is much larger than we have dared to suppose. This means that most cosmological observations in the past have occurred in the "noise" of the sub-Hubble sphere. Also, it means that many of the approximations used to project the world picture onto the world map, which are valid only for small redshifts, are no longer valid when applied to observations of higher redshift beyond the sub-Hubble sphere. We are thus caught between the devil and the deep blue sea. This is a crazy idea, but crazy ideas are not always wrong.

References and detailed discussions can be found in the author's *Cosmology: The Science of the Universe*, 2nd edition, Cambridge University Press, 2000.

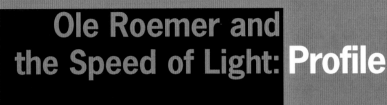

Ole Roemer and the Speed of Light: Profile

Ole Roemer at his transit telescope in Copenhagen.

In 1676, the Danish astronomer Ole Roemer (1644–1710) became the first person to measure the speed of light. Until that time, scientists assumed that the speed of light was either too fast to measure or infinite. The dominant view, vigorously argued by the French philosopher Descartes, favored an infinite speed.

Roemer, working at the Paris Observatory, was not looking for the speed of light when he found it. Instead, he was compiling extensive observations of the orbit of Io, the innermost of the four big satellites of Jupiter discovered by Galileo in 1610. By timing the eclipses of Io by Jupiter, Roemer hoped to determine a more accurate value for the satellite's orbital period. Such observations had a practical importance in the seventeenth century. Galileo himself had suggested that tables of the orbital motion of

Jupiter's satellites would provide a kind of "clock" in the sky. Navigators and mapmakers anywhere in the world might use this clock to read the absolute time (the standard time at a place of known longitude, like the Paris Observatory). Then, by determining the local solar time, they could calculate their longitude from the time difference. This method of finding longitude eventually turned out to be impractical and was abandoned after the development of accurate seagoing timepieces. But the Io eclipse data unexpectedly solved another important scientific problem—the speed of light.

The orbital period of Io is now known to be 1.769 Earth days. The satellite is eclipsed by Jupiter once every orbit, as seen from the Earth. By timing these eclipses over many years, Roemer noticed something peculiar.

The time interval between successive eclipses became steadily shorter as the Earth in its orbit moved toward Jupiter and became steadily longer as the Earth moved away from Jupiter. These differences accumulated. From his data, Roemer estimated that when the Earth was nearest to Jupiter (at E1 in the Figure 1), eclipses of Io would occur about eleven minutes earlier than predicted based on the average orbital period over many years. And 6.5 months later, when the Earth was farthest from Jupiter (at E2), the eclipses would occur about eleven minutes later than predicted.

Roemer knew that the true orbital period of Io could have nothing to do with the relative positions of the Earth and Jupiter. In a brilliant insight, he realized that the time difference must be due to the finite speed of light. That is, light from the Jupiter system has to travel farther to reach the Earth when the two planets are on opposite sides of the Sun than when they are closer together. Romer estimated that light required twenty-two minutes to cross the diameter of the Earth's orbit. The speed of light could then be found by dividing the diameter of the Earth's orbit by the time difference.

The Dutch scientist Christiaan Huygens, who first did the arithmetic, found a value for the speed of light equivalent to 131,000 miles per second. The correct value is 186,000 miles per second. The difference was due to errors in Roemer's estimate for the maximum time delay (the correct value is 16.7, not 22 minutes), and also to an imprecise knowledge of the Earth's orbital diameter. More important than the exact answer, however, was the fact that Roemer's data provided the first quantitative estimate for the speed of light, and it was in the right ballpark.

Roemer returned to Denmark in 1681, where he pursued a distinguished career in both science and government. He designed and built the most accurate astronomical instruments of his time and made extensive observations. He later served as mayor and prefect of police of Copenhagen and ultimately as head of the State Council. Roemer is remembered today of course not for his high political office but for being the first person to measure the speed of light.

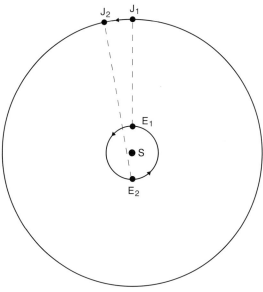

Orbits of Jupiter and Earth

Roemer measured the speed of light by timing eclipses of Jupiter's moon Io. In this figure, S is the Sun, E1 is the Earth when closest to Jupiter (J1) and E2 is the Earth about six months later, on the opposite side of the Sun from Jupiter (J2). When the Earth is at E2, the light from the Jupiter system has to travel an extra distance represented by the diameter of the Earth's orbit. This causes a delay in the timing of the eclipses. Roemer measured the delay and, knowing approximately the diameter of the Earth's orbit, made the first good estimate of the speed of light.

Olbers' Paradox:
Why Is the Sky Dark at Night?

The oldest and simplest astronomical observation tells us something profound about the universe. The sky is dark at night. It isn't obvious why this should be so. If you stand in a small grove of trees and look toward the horizon, you can see patches of sky in the distance between the tree trunks. But if you stand in a large forest, your view is everywhere blocked by a "solid wall" of tree trunks. Extending the analogy to three dimensions, if the universe of stars is large enough, your line of sight should be blocked in every direction by a "solid wall" of stars. If you could magnify that view sufficiently, the sky would everywhere look something like the image on the next page.

The entire sky would be about as bright, and as hot, as the surface of the Sun. The immense distance to the stars making up the "wall of light" would have no effect on the total amount of energy reaching us. We should be surrounded by a blazing oven of light. Instead the night sky is practically black. So where does the argument go wrong?

The German astronomer Johannes Kepler first posed this problem in 1610. He also suggested a solution: the universe of stars, he believed, extends only out to a finite distance; once your line of sight passes that boundary, it encounters only empty space. But how far is that boundary? Why is it there? And what lies beyond it?

Astronomers after Kepler proposed various solutions to the problem of the dark night sky, which came to be called Olbers' Paradox. In 1823, the German astronomer Heinrich Olbers suggested that starlight is gradually absorbed while traveling through space, and this cuts off the light from any stars beyond a sufficiently

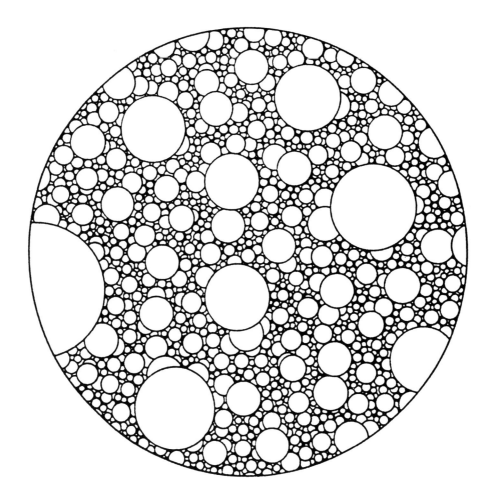

A "forest" of stars.

great distance. But that doesn't solve the problem, either. Any absorbing interstellar gas or dust would simply heat up until it reradiated all the starlight it absorbed, and the energy reaching us would be the same. By analogy, sprinkling the air in a hot oven with absorbing dust won't cool it for very long.

So why is the night sky dark? The first scientifically reasonable answer was given in 1848 by the American poet and writer Edgar Allan Poe! He suggested that the universe is not old enough to fill the sky with light. The universe may be infinite in size, he thought, but there hasn't been enough time since the universe began for starlight, traveling at the speed of light, to reach us from the farthest reaches of space.

Astronomers have concluded that the universe began some 12 to 15 billion years ago. That means we can only see the part of it that lies within 12 to 15 billion light-years from us. There may be an infinite number of stars beyond that cosmic horizon but we can't see them because their light has not yet arrived. And the observable part of the universe contains too few stars to fill up the sky with light.

But that is not the whole solution to the paradox. Most stars, like the Sun, shine for a few billion years or so before they consume their nuclear fuel and grow dark. Dying stars spew gas and dust back into space, and this material gives birth to new generations of stars. But after enough generations, all the nuclear fuel in the universe is eventually exhausted, and the formation of luminous stars must come to an end. So even if the universe were infinitely old as well as infinitely large, it would not contain enough fuel to keep the stars shining forever and to fill up all of space with starlight. And so the night sky is dark.

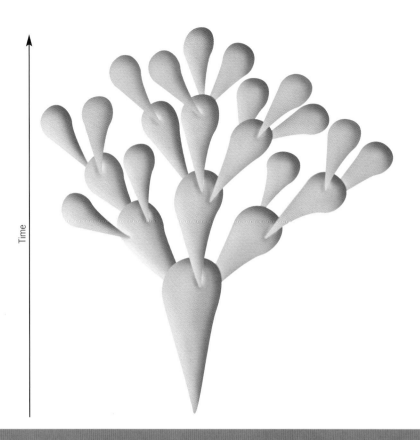

Time

What Happened Before the Big Bang?

Lee Smolin

What was the universe like before the Big Bang? This may seem like an absurd question, for was not the Big Bang the beginning of the universe? In popular treatments, the term is often used in that sense. In a scientific discussion, however, we must be more careful.

The Big Bang theory was suggested by the observation that the distant galaxies are moving apart from one another. The universe is expanding, and therefore must have been denser in the past.

Lee Smolin is Professor of Physics in the Center for Gravitational Physics and Geometry, Penn State University.

Like a gas, a denser universe was also hotter. Observations with microwave telescopes show that the very early universe was indeed hotter. We can see a steady glow of microwave energy coming from all parts of the sky. This so-called cosmic microwave background radiation is the relic of a time, about half a million years after the Big Bang, when the universe was everywhere almost as hot as the surface of a star. Using nuclear physics, we can trace the universe back even earlier, to a time when it must have been as hot as the center of a star, and nuclear reactions were making most of the helium in the universe.

We can extrapolate before that, when the universe was even denser and hotter. There is good reason to think that, at some point extremely early in its history, the universe expanded exponentially from a state some ninety orders of magnitude denser than nuclear matter. If we were naive, we might conclude that just before that period of so-called "inflation," the universe was infinitely dense. But Einstein's Theory of General Relativity, our best description of space and time, breaks down when applied to regions of infinite density, and the concept of time vanishes. So general relativity cannot describe what, if anything, happened before that point.

To describe such extreme conditions, we need a quantum theory of gravity. Such a theory would unify quantum theory, which describes the behavior of subatomic particles, and general relativity, which describes gravity as the curvature of space by matter. A great deal has been learned in recent years about a quantum theory of gravity, although our knowledge of it is still far from complete. What we know suggests that there is a limit to how dense a region of the universe can get before the

picture of continuous space breaks down, to be replaced by some kind of quantized network of causal relationships. At this level there is no space, but only a grainy structure that has the same relationship to continuous space as the atoms that make up a sheet of gold have to their smooth outward appearance. But even under such extreme conditions, the theory tells us, time may still exist. This leaves us with no scientific reason to conclude that time itself began in the Big Bang.

Let's turn then to a second, seemingly unrelated question: What happens to a massive star when it collapses to form a black hole? Stars are held up against gravity by the thermal energy they generate in nuclear reactions. Stars collapse when they exhaust their available nuclear fuel. Most collapsed stars settle down into compact, Earth-sized objects called white dwarfs. But the more massive stars, those greater than about eight times the mass of the Sun, become unstable at this point and explode. We call this a supernova. Most of the mass is expelled into interstellar space, leaving a collapsing core. If this remnant has less than a certain critical mass, it will form a stable neutron star—an object about fifteen miles across with the density of an atomic nucleus. But there is a limit to the mass of a stellar remnant, above which nothing can stop the collapse. A more massive object collapses past the point where gravity is strong enough to prevent the escape of light. From then on it is a black hole.

We have good evidence that black holes indeed exist. There are perhaps a billion of them in the Milky Way Galaxy alone, with new ones being born, somewhere in the observable universe, at least ten times per second. What happens to all those collapsing stars after they become black holes? General Relativity predicts that they become infinitely dense. The process of collapse into a black hole thus resembles the

Big Bang in reverse. But as we have seen, the Theory of General Relativity breaks down at infinite density. It does not take quantum physics into account, and so cannot be trusted there.

One possibility is that at some extreme density the structure of the vacuum—in a certain sense space itself—undergoes a change of phase, somewhat analogous to the sudden change of a material substance from a solid to a liquid. This results from a rearrangement of certain fields, analogous to the electric and magnetic fields, that are known to exist in the vacuum. Such fields determine the observed masses of the elementary particles. Normally those fields are constant, which is why all electrons, for example, have the same mass. But under conditions of extreme density, as in the center of a black hole, the fields begin to vary, resulting in a kind of "melting" of the vacuum. The laws governing nuclear reactions actually predict that such a transition occurs. According to the theory, that is exactly what happened early in the history of our universe, causing the brief burst of exponential expansion called "inflation." If such conditions also occur inside a black hole, the region of space occupied by the collapsing star will begin to expand rapidly again. It may become in time as large as our own universe.

But wouldn't we notice if new universes were bursting into existence all around us? The point is that we cannot see it, because it is happening inside a black hole, and no light or any kind of information can leave that region. Even if the interior blows up to become as big as the universe itself, no light escapes from that region to ourselves. This may seem absurd. How can a black hole produce something as big as our universe if it forms inside our universe? This is only absurd if one thinks of the geometry of space as invariable. What General Relativity tells us, as confirmed by numerous precise

observations, is that the geometry of space is a dynamical quantity, which can vary without limit. The formation of a huge region inside a black hole, completely cut off from any possibility of being observed by us, is perfectly compatible with everything we know about the laws of physics.

But where does the energy come from to create a whole universe inside a black hole? The surprising answer is that it doesn't take much energy at all to create a universe. That's because the energy in the gravitational fields created by and surrounding the matter is always negative, and this cancels out the positive energy residing in the matter itself. With no net change in energy, the emergence of a universe would not be observed gravitationally outside the black hole.

Both kinds of exponential expansions, that which occurred in our early inflationary universe and those that may happen inside black holes, are conjectural—they have not yet been confirmed observationally. However, at least in the first case a confirmation may be possible. Some consequences of the inflationary model of the early universe will soon be tested by a satellite that will make detailed measurements of the cosmic microwave background.

So let us just take one more step: what if the two kinds of events are essentially the same? What if the Big Bang that launched the expansion of our universe began with a brief era of inflation? And what if a new Big Bang occurs at the center of every black hole that forms within our universe? In that case, could our own universe have been born in a black hole, formed by the death of a star in a different universe?

If the answer is yes, then our own universe is just one of an enormous number of similar universes, each one born from a previous one, and many of them giving birth to new ones. This sounds like an absurd fantasy. But could it be true? More to the point, is this hypothesis testable? Are there possible observations that might prove it false? (No scientific hypothesis is taken seriously unless there is some way to test it by observation.) And, most importantly, does it offer a chance at explaining something that otherwise would remain mysterious?

The answer to all of these questions is yes! The reason is that all the properties of the elementary particles, such as their masses and the strengths of the forces between them, are determined by the values of certain fields in the vacuum. These fields are all "frozen" presently, so all these quantities are unchanging. But in a phase transition that triggers a burst of inflationary expansion, they become unfrozen. During inflation the universe cools rapidly, and the fields "freeze" again. But when they do, their values may have changed. The result is that the masses of the elementary particles and the strengths of the physical forces, the so-called parameters of our universe, may be different in each new universe.

According to our best theories of the elementary particles—string theories and grand unified theories—these changes must take place. But it is natural to suppose that they are small, since the cooling during an inflationary phase occurs so rapidly that the parameters will be frozen again before they have time to change very much. In that case, we can draw some remarkable predictions from this scenario.

The reason, surprisingly, is that we can now apply to cosmology the methods of mathematical genetics. The collection of universes resembles a growing population of bacteria. The particle masses and other parameters that govern the properties of a universe are like its genes. They vary slightly in each generation,

just as do the genes of bacteria. We then have an exponentially growing population of universes, with each generation inheriting the traits of its parents, but slightly modified by random mutations. The process of natural selection will then sort out which combinations of genes, or parameters, will be most common. The answer is the ones that most successfully reproduce.

This is true for bacteria. Might it apply to the universe itself? Let us follow the argument and see where it goes. Suppose some combinations of parameters result in universes that produce more of the massive stars that make black holes than other combinations. After a few "generations" almost every universe will be one of these, not because the others are absent, but because they will become so vastly outnumbered. A universe, chosen at random, is much more likely to have a parent with many progeny, and most universes will resemble their parents.

If our universe is typical, then it also has this property of approaching maximum reproductive success. This means that the parameters that govern our universe have been fine-tuned by natural selection to maximize the number of black holes. This leads to the remarkable prediction that any change in the parameters from their present values should lead to a universe that produces fewer black holes.

Can this be checked? Obviously we cannot do an experiment to vary the parameters of a universe and see if it produces fewer black holes. But we can try to deduce, from our knowledge of astrophysics, whether it is really the case that all changes in the existing parameters lead to a universe with fewer black holes.

Here is the situation at present: our standard theory of elementary particle physics has about twenty parameters. There are good arguments that increases *and* decreases in five of them would lead to a world with fewer black holes. These five are the ratio of the electron mass to the proton mass and the strengths of the four fundamental forces of physics: electromagnetism, the strong and weak nuclear forces, and the gravitational force. There are three other parameters for which an increase must lead to a universe with fewer black holes, while a decrease either has no effect or it is not possible to tell with present knowledge. For the remaining parameters, there is either no strong effect, or it is so far not possible to deduce the effect.

These results say that the number of black holes produced depends sensitively on the laws of physics. Stars are made in the collapse of interstellar gas clouds. In order to form stars massive enough to become black holes, such clouds must cool rapidly. Otherwise, their heat would drive away the gas before it could assemble a massive star. In our universe much of this cooling is accomplished by the element carbon, which efficiently radiates away heat from a collapsing cloud. But the nuclear reactions that make carbon are uniquely sensitive to the parameters of physics. Thus, we living things share something with black holes: the world is hospitable to us because it is full of carbon.

There are a few possible observations that would rule out this hypothesis. One of them would be the observation of any neutron star with a mass more than twice that of the Sun. To maximize the number of black holes, the hypothesis predicts that the upper limit for the mass of neutron stars (above which they are unstable and become black holes) should be as low as possible. If there is a parameter which when changed can lower that limit, its value must be optimized by natural selection. It turns out, according to a remarkable theory of Hans

Bethe and Gerald Brown, that the mass of the strange quark, one of the elementary particles, should be such a parameter. According to their theory, the upper limit for a neutron star can be lowered to about 1.5 times the mass of the Sun by adjusting that parameter. This is just above the mass of the heaviest neutron star so far seen. One observation of a neutron star sufficiently above this value and the story I have been telling here is wrong.

A second observation that could kill this theory would be one showing that more black holes were formed early in the universe, before carbon was abundant, than are being formed today. This could be accomplished by counting supernovas at great distances (and hence far back in time) with the Next Generation Space Telescope.

Whether this theory is right or wrong, it points to an important lesson. There is an old prejudice that it is more scientific to explain things in terms of absolute principles than in terms of contingent, historical events. The biologists learned a long time ago that this is a myth. As the American philosopher Charles Pierce said a century ago, there is nothing more in need of explanation than an absolute principle or law. Ever since Darwin, biology has flourished because the biologists discovered that all they need of absolute principles are a few facts about how random statistical processes operate. The mechanisms of self-organization of non-equilibrium systems, chief of which is natural selection itself, were sufficient to make the biological world, and they seem to be all we need to understand it. True or false, the story I have told here suggests that the same may be true for the laws of physics and the origin and history of our universe. 🪐

Georges Lemaître, Father of the Big Bang: Profile

According to the Big Bang theory, the expansion of the observable universe began with the explosion of a single particle at a definite point in time. This startling idea first appeared in scientific form in 1931, in a paper by Georges Lemaître, a Belgian cosmologist and Catholic priest. The theory, accepted by nearly all astronomers today, was a radical departure from scientific orthodoxy in the 1930s. Many astronomers at the time were still uncomfortable with the idea that the universe is expanding. That the entire observable universe of galaxies began with a bang seemed preposterous.

Lemaître was born in 1894 in Charleroi, Belgium. As a young man he was attracted to both science and theology, but World War I interrupted his studies (he served as an artillery officer and witnessed the first poison gas attack in history). After the war, Lemaître studied theoretical physics, and in 1923 was ordained as an abbé. The following year, he pursued his scientific studies with the distinguished English astronomer Arthur Eddington, who regarded him as "a very brilliant student, wonderfully quick and clear-sighted, and of great mathematical ability." Lemaître then went on to America, where he visited most of the major centers of astronomical research. Later, he received his Ph.D. in physics from the Massachusetts Institute of Technology.

In 1925, at age 31, Lemaître accepted a professorship at the Catholic University of Louvain, near Brussels, a position he retained through World War II (when he was injured in the accidental bombing of his home by U.S. forces). He was a devoted teacher who enjoyed the company of students, but he preferred to work alone. Lemaître's religious interests remained as important to him as science throughout his life, and he served as President of the Pontifical Academy of Sciences from 1960 until his death in 1966.

In 1927, Lemaître published in Belgium a virtually unnoticed paper that provided a compelling solution to the equations of General Relativity for the case of an expanding universe. His solution had, in fact, already been derived without his knowledge by the Russian Alexander Friedmann in 1922. But Friedmann was principally interested in the mathematics of a range of idealized solutions (including expanding and contracting universes) and did not pursue the possibility that one of them might actually describe the physical universe. In contrast, Lemaître attacked the problem of cosmology from a thoroughly physical point of view, and realized that his solution predicted the expansion of the real universe of galaxies that observations were only then beginning to suggest.

By 1930, other cosmologists, including Eddington, Willem de Sitter, and Einstein, had concluded that the static (non-evolving) models of the universe they had worked on for many years were unsatisfactory. Furthermore, Edwin Hubble, using the world's largest telescope at Mt. Wilson in California, had shown that the distant galaxies all appeared to be receding from us at speeds proportional to their distances. It was at this point that Lemaître drew Eddington's attention to his earlier work, in which he had derived and explained the relation between the distance and the recession velocity of galaxies. Eddington at once called the attention of other cosmologists to Lemaître's 1927 paper and arranged for the publication of an English translation. Together with Hubble's observations, Lemaître's paper convinced the majority of astronomers that the universe was indeed expanding, and this revolutionized the study of cosmology.

A year later, Lemaître explored the logical consequences of an expanding universe and boldly proposed that it must have originated at a finite point in time. If the universe is expanding, he reasoned, it was smaller in the past, and extrapolation back in time should lead to an epoch when all the matter in the universe was packed together in an extremely dense state. Appealing to the new quantum theory of matter, Lemaître argued that the physical universe was initially a single particle—the "primeval atom" as he called it—which

disintegrated in an explosion, giving rise to space and time and the expansion of the universe that continues to this day. This idea marked the birth of what we now know as Big Bang cosmology.

It is tempting to think that Lemaître's deeply-held religious beliefs might have led him to the notion of a beginning of time. After all, the Judeo-Christian tradition had propagated a similar idea for millennia. Yet Lemaître clearly insisted that there was neither a connection nor a conflict between his religion and his science. Rather he kept them entirely separate, treating them as different, parallel interpretations of the world, both of which he believed with personal conviction. Indeed, when Pope Pius XII referred to the new theory of the origin of the universe as a scientific validation of the Catholic faith, Lemaître was rather alarmed. Delicately, for that was his way, he tried to separate the two:

"As far as I can see, such a theory remains entirely outside any metaphysical or religious question. It leaves the materialist free to deny any transcendental Being… For the believer, it removes any attempt at familiarity with God… It is consonant with Isaiah speaking of the hidden God, hidden even in the beginning of the universe."

In the latter part of his life, Lemaître turned his attention to other areas of astronomical research, including pioneering work in electronic computation for astrophysical problems. His idea that the universe had an explosive birth was developed much further by other cosmologists, including George Gamow, to become the modern Big Bang theory. While contemporary views of the early universe differ in many respects from Lemaître's "primordial atom," his work had nevertheless opened the way. Shortly before his death, Lemaître learned that Arno Penzias and Robert Wilson had discovered the cosmic microwave background radiation, the first and still most important observational evidence in support of the Big Bang.

Section Five: Life in the Universe

The Horsehead Nebula in Orion is a giant molecular cloud. Such dark interstellar clouds are rich in the substances needed to form stars, planets, and life. Kitt Peak National Observatory.

Introduction Steven Soter

Scientists had long assumed that the Earth's deep sea floor, secluded in perpetual darkness, was a biological desert. We had to abandon that notion in 1977, after marine geologists descended more than two kilometers to investigate a chain of volcanic vents on the sea floor near the Galapagos Islands. The searchlights from their submersible vehicle pierced the darkness to reveal a rich profusion of strange and unknown animals flourishing in the shimmering hot waters spewing from volcanic vents.

Here was an unsuspected realm of life on Earth. The deep sea vent communities depend on heat-tolerant bacteria that form the base of their food chain. These bacteria obtain energy and nutrients from the Earth itself, by exploiting chemical reactions between the hot volcanic gases and the crustal rocks and sea water.

The discovery of these ecosystems overturned the long-established dogma that all life on Earth depends directly or indirectly on solar energy via photosynthesis. It would also lead to a revolution in our thinking about life in the universe.

All known life forms require liquid water. Many scientists had assumed that the Earth was the only possible abode of life in our solar system because it alone has the right atmospheric temperature and pressure to allow liquid water on the surface. But our conception of what constitutes a "habitable zone" has expanded with the discovery of ecosystems that are independent of sunlight and with the growing realization that liquid water probably exists below the surfaces of many other worlds.

There is growing evidence for the existence of an extensive "deep hot biosphere" in the fluid-filled cracks of the Earth's outer crust. Microbial life has been verified in deep bore holes drilled in many parts of the world. The microbial communities of the deep sea vents appear to represent merely the exposed tip of a far more extensive ecosystem hidden in the outer crust of the planet. This realm of life may extend as deep as five to ten kilometers, depending on the geological setting. Below those depths, increasing temperatures will destroy proteins. But the volume of fluid-filled fractures down to such depths is enormous. The total biomass of the dark subsurface life on Earth may be comparable to that of our familiar surface biosphere.

Water can exist in liquid form only within a restricted range of temperatures and pressures. The Earth is unique in our solar system in having just those conditions on its surface. Liquid water is more likely to occur below the surface of a world, because pressure and temperature both increase with depth. A large

A volcanic sulfide chimney ("black smoker") on the deep dark seafloor supports an ecosystem that uses chemical energy flowing from the Earth's interior. Such ecosystems show that life does not require light.

rocky planet can generate enough heat by its natural radioactivity to maintain water in the liquid state over a wide range of depths. Tidal energy may strongly heat the interiors of large satellites in suitable orbits. The temperatures a few kilometers below the surfaces of many worlds are dominated by such energy sources and are relatively insensitive to the solar contribution. Subsurface liquid water may therefore be a relatively common occurrence in the universe. In contrast, liquid water on the surface may be the rare exception.

The atmosphere of an earthlike planet is a relatively delicate structure, vulnerable to powerful outside influences. During its first half billion years, the Earth's surface was subjected to extreme climate fluctuations, including catastrophic events triggered by the impacts of massive meteorites. But the subsurface environment would have been a more stable place. Perhaps life on Earth began in those sheltered depths, and only later worked its way to the surface, evolving photosynthesis and the other adaptations required for the more demanding surface environment.

The "family tree" of life on Earth, deduced from comparative genetic studies of all known living groups, strongly suggests that the common ancestral life forms were heat-tolerant microbes. This finding is consistent with the hypothesis that life originated in the dark hot region below our planet's surface. If this is the case, it may have done so on many other planets as well.

The discovery of Martian meteorites on Earth set in motion another major change in our thinking about life in the universe. Many scientists had long assumed that the planets of our solar system are biologically isolated. But this now seems unlikely. When a comet or asteroid collides with a solid body like Mars or the Earth, some of the debris from the resulting explosion crater escapes from the planet's gravity. Some of the escaping material will be in the form of rocks and boulders that manage to avoid being strongly shocked or pulverized. Many of these eventually reach other planets.

Rocks ejected from Mars are landing on Earth, while rocks from the Earth are falling on Mars and on the other worlds in our solar system. Many of the rocks blown into space from the Earth's outer crust will contain living microbes. Laboratory experiments show that certain terrestrial microbes can survive the shock of ejection and the frozen vacuum and radiation of interplanetary space for an indefinite time. So the natural transport of life between worlds in a planetary system appears to be a distinct possibility. If so, a single origin of life on one world may be sufficient to spread life to all the other suitable worlds in that system.

Perhaps the deliberate seeding of life on sterile worlds will eventually become possible. Mars is now a frozen desert, but its surface shows signs that liquid water occasionally bursts out of the ground and erodes gullies before evaporating. There is also evidence that liquid water was stable on the Martian surface a few billion years ago under a denser and warmer atmosphere. Much of the ancient Martian water and atmosphere is evidently still trapped in the polar ice caps and below the ground. Perhaps they could be gradually released by human engineering on a planetary scale, making the surface of the planet habitable. Carefully selected species could then be introduced. Even if it turned out to be possible, such "terraforming" of Mars would probably take many thousands of years. But the theoretical study of the problem raises interesting scientific and ethical questions.

To explore the prospects for life in the universe, we pose the following questions:

What makes a world habitable?

Christopher Chyba, holder of the Carl Sagan Chair for the Study of Life in the Universe at the SETI Institute and Associate Professor in the Department of Geological and Environmental Sciences at Stanford University, shows how recent discoveries have expanded the possible range of environments for life in the universe.

How can life exist without energy from sunlight?

Thomas Gold, Professor Emeritus of Astronomy at Cornell University, outlines a general theory of deep subsurface life on the Earth and in the universe. This essay is adapted from his landmark 1992 paper which first proposed the existence of an extensive "deep hot biosphere" in the fluid-filled cracks of the Earth's outer crust.

Can life spread naturally across interplanetary space?

H. J. Melosh, Professor of Planerary Science at the Lunar and Planetary Laboratory at the University of Arizona in Tucson, explains how boulders ejected from impact craters escape intact from planets, and how microbes inside them can survive the rigors of interplanetary space. Could life thereby be transported to "inoculate" other worlds?

What is "terraforming"?

Christopher P. McKay, Research Scientist at the NASA Ames Research Center in Mountain View, California, examines how our descendants might be able to make Mars habitable, and why it won't be easy.

Earthrise over moon. The Earth is flourishing with life,
while the nearby Moon is a dead dry cinder. The difference
is mainly due to liquid water. This image was taken by
the Apollo 11 astronauts in 1969.

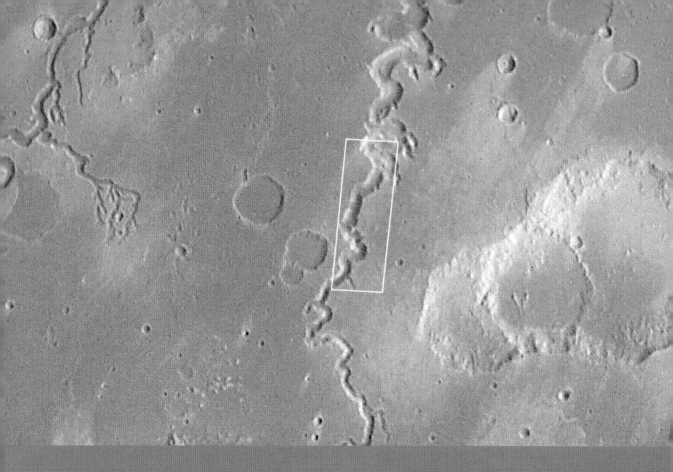

Habitable Worlds

Christopher Chyba

What makes a world habitable for life? Whether or not we regard a planet or a moon as habitable depends on who or what we imagine might be doing the inhabiting. Before the rise of oxygen in the Earth's atmosphere around 2.5 billion years ago, the Earth was uninhabitable for human beings as well as plants, animals, and fungi—all of which require molecular oxygen (O_2). But prior to this time, the Earth was certainly habitable for a wide variety of highly successful microorganisms. Indeed, it was photosynthesizing bacteria that probably produced the oxygen that made Earth habitable for higher organisms.

Christopher Chyba is Carl Sagan Chair for the Study of Life in the Universe, SETI Institute; Associate Professor (Research) Department of Geological and Environmental Sciences, Stanford University; and Co-Director of the Center for International Security and Cooperation, Stanford University.

Evidence for ancient water on the Martian surface.
Part of a meandering canyon in the Xanthe Terra region.
The valley is about 2.5 kilometers wide and has a small
channel at the bottom, which appears to have been formed
by erosion from water flowing over an extended period of
time. Area in rectangle at left is shown enlarged at right.
Image obtained by the Mars Global Surveyor.

160 161

Ancient fossils demonstrate that microbial life existed on Earth for billions of years before multicellular fungi, plants, or animals appeared. So if we want to cast the broadest net for life, we should probably choose a definition of "habitability" appropriate to the simplest organisms that we would judge to be alive. But how do we make this judgment?

Finding a satisfactory general definition of life has proven elusive. As a practical matter, it may be better to focus on those conditions that are necessary for life as we know it: liquid water, a source of carbon (and other key elements), and a source of usable energy. Since energy from the Sun is available throughout much of the solar system, and since carbon, often in the form of organic molecules, is also common, the search for liquid water has for many years been the focus in the search for life beyond Earth. In fact, over the last three decades, the usage of the word "habitability" shifted from meaning those conditions needed to support human life, to the less stringent conditions necessary for the stability of liquid water at a world's surface. Recently, as we have learned more about the "deep biosphere" of microorganisms in the Earth's subsurface, the definition is widening once again.

The existence of liquid water at a world's surface no longer seems likely to be a requirement for habitability by all forms of life. Recent estimates of the mass of the deep biosphere on Earth suggest that it may be comparable in total biomass to that of the surface biosphere (see the essay by Thomas Gold in this section). This is a remarkably different image than the traditional picture of a planet whose biomass is dominated by forests of trees.

A critical question for the exobiological implications of the deep terrestrial biosphere is the extent to which subsurface ecosystems are truly independent of surface life and conditions. Despite first appearances, much of that biosphere is not ultimately independent of the surface, because it makes use directly or indirectly of the oxygen or organic molecules produced in abundance by photosynthesizing surface life. It seems likely, however, that some organisms in the subsurface biosphere (certain methane-producing bacteria, for example) could continue to exist over geological time even if our Sun were somehow extinguished and the surface of the Earth frozen over permanently.

If this is right, then by analogy life might even today persist in subsurface liquid water habitats on Mars, having retreated to those niches as the Martian surface became inhospitable billions of years ago. Spacecraft images of the oldest Martian terrain reveal networks of now-dry valleys that were apparently cut by running water, as well as numerous lakes where some of this water once pooled. But after about 3.5 billion years ago, the Martian surface seems to have become a frozen desert.

Another crucial question is whether life can originate in the absence of the abundant energy that can be harvested from the Sun. On Mars, life could have begun at the surface, and only later been forced to become an exclusively subsurface phenomenon. But on Jupiter's ice-covered moon Europa, if the origin of life did occur, it probably took place far below the reach of sunlight. While it is not yet certain that Europa has an ocean, a number of independent lines of evidence, mostly from observations by the Galileo spacecraft, now point in that direction (see the essay by Clark R. Chapman in Section One). Europa may well harbor the second ocean in the solar system, hidden beneath kilometers of ice. If the origin of life cannot occur without the abundant energy source of the Sun, then any subsurface ocean of Europa—however hospitable it might prove to be to some forms of life that exist on Earth today—may have remained sterile for the past four billion years. Such a discovery, while less important than finding life in the oceans of Europa, would be nearly as startling.

Solar system exploration over the coming decade should go a long way towards answering the questions of whether life might have once existed, or may still exist, on Mars or Europa. Between now and 2010, NASA is planning an ambitious program of Martian exploration. Around the end of the decade, these missions will include the return to Earth of carefully selected Martian samples for examination in terrestrial laboratories. In addition, a series of Mars "micromissions" will begin to establish the infrastructure for a more sophisticated presence on the Red Planet. By 2010 we should have established a Mars global positioning system and internet. As the internet becomes interplanetary, citizens of Earth will be able to view live video returned from robots exploring Mars by ground and air. Millions of people will visit Mars regularly in this virtual sense, and Mars will come to seem familiar to them in the same way that the Grand Canyon or Mt. Everest is today. Over the next few decades, this familiarity will inexorably lead us into thinking of ourselves as a civilization that naturally spans the solar system.

The most promising sites for current life on Mars—for example, hot spring environments or deep aquifers, if either should exist—may well be the most demanding for robot explorers. In the end, we will send humans to explore the Red Planet. Working together with robots, humans will ultimately explore in detail, or drill down to, those sites identified as the most likely remaining locations for life.

If there is life on Mars, we will learn whether it shares a common ancestor with life on Earth.

An important realization of the past ten years is that the worlds of the inner solar system have not necessarily been biologically isolated; viable organisms appear able to move between Earth and Mars by being enclosed in surface rocks "spalled off" by large impacts (see the essay by H. J. Melosh in this section). It is at least possible that whichever world developed life first then "inoculated" the others. If life still exists on Mars, it could be that we share a common ancestor with it. If so, we should be able, on the basis of molecular similarities (e.g., by comparison of DNA sequences), to demonstrate that this is the case. Of course, should Martian life exist and be of independent origin from life on Earth, it might lack DNA altogether.

Europa may be the most promising site for life elsewhere in the solar system, but it will be far more demanding than Mars to explore. Direct spacecraft trajectories to Jupiter take three years, and require much more energy, than do flights to Mars. Moreover, Europa resides deep within Jupiter's strong magnetosphere, which bathes Europa's surface in charged particle radiation. This radiation would be of little concern to any Europan organisms living beneath kilometers of ice, but it is extremely demanding on a spacecraft's electronic components. Nevertheless, the exploration of Europa will begin with an orbiter mission planned to launch in 2003, and designed to answer definitively whether or not there is a Europan ocean. If the answer is yes, as seems increasingly likely, there will be a program of exploration, including landers, designed to examine the geology, chemistry, and possible biology of Europa.

There are hints from the Galileo spacecraft now orbiting Jupiter that subsurface oceans may in fact be typical for large icy satellites such as Jupiter's moon Callisto and Saturn's moon Titan.

Because Titan is covered with an atmospheric smog layer, we have not yet seen its surface in any detail. But in 2004, the Huygens probe will drop into its atmosphere, float down by parachute over two hours, and return images to the Earth. Some scientists have suggested that there may be liquid hydrocarbons flowing on Titan's surface. If Titan is a world with hydrocarbon weather, how Earth-like—yet how alien—it will seem! And if Titan's surface organics make contact with subsurface liquid water, what else might be possible there?

One challenge in our exploration of all these worlds will be planetary protection. NASA has guidelines, rooted in international treaty, to protect the worlds that it visits against possible "forward contamination" by microorganisms carried along in spacecraft from Earth. We have further progress to make in reducing the "bioload" of missions we launch to worlds such as Europa—progress demanded legally by international treaty, scientifically by the requirement of not introducing organisms that could be confused with indigenous ones, and ethically by an environmental perspective that recognizes the requirement to protect alien biospheres as well as our own.

What awaits us beyond our solar system? One of the most remarkable scientific results of the past decade has been the discovery of more planets in orbit around other suns than around our own (see the essays by David C. Black and Steven V. W. Beckwith, in Sections One and Two, respectively). We will soon have catalogs of extrasolar planetary systems analogous to those we now have for stars. We will then know whether our particular planetary system is typical or unusual—although the results in hand so far suggest that ours will turn out to be something in between.

Currently the only extrasolar worlds our technologies routinely detect are giant planets about the size of Jupiter. We cannot yet reliably detect other worlds the size of Earth, although attempts are being made even as this essay is written. But the extrasolar giant planets we do find could have large icy satellites similar to those of the gas giants in our own solar system. Some of these, in turn, could support subsurface liquid water oceans. For that matter, those extrasolar giant planets that are at the right distance from their central star might even have icy satellites where liquid water is stable at the surface.

Our continuing exploration of the planets, either within or beyond our own solar system, will help us understand the environmental context for life on Earth. For example, the intense bombardment of the Earth by comets and asteroids in the first 800 million years of its history was discovered by deciphering the cratering history of the Moon. This ancient cratering record had long since been erased from Earth's surface, because Earth is so geologically active and terrestrial erosion is so relentless.

In the same way, our investigations of the enormous greenhouse effect on Venus and the failed greenhouse on Mars have tested our ideas about Earth's climate, and the feedback cycles that maintain our planet's surface temperatures within the right range for liquid water to exist. Although we fear a greenhouse effect on Earth that could become too strong, the right amount of greenhouse is essential: without the warming blanket of water vapor, carbon dioxide, and other gases in our atmosphere that provide us with over thirty degrees Centigrade of greenhouse warming, the surface of the Earth would be completely frozen over year round, and our planet would resemble Europa.

On Earth, there seems to be a long-term negative feedback cycle for carbon dioxide that keeps greenhouse warming within the temperature limits required for liquid water. But Mars and Venus provide examples of worlds where this sort of balance broke down, with disastrous consequences. Venus had a runaway greenhouse effect that led to a dense carbon-dioxide atmosphere and surface temperatures hot enough to melt lead. The Martian greenhouse failed in the opposite direction, leaving a wispy-thin carbon-dioxide atmosphere and a frozen desert world. Liquid water is not stable on the surface of either planet.

The bombardment history of the Moon has proven critical to our thinking about the environment of the early Earth and the origin and evolution of life on this planet. Our growing knowledge of the atmospheric evolution of Venus and Mars has helped us understand the importance of the stability of greenhouse warming for maintaining equable temperatures at a planet's surface. The exploration of other worlds has become part of our search for understanding and maintaining the habitability of our own.

Looking farther ahead, how will the solar system appear to us at the midpoint of the twenty-first century? We should know by then if Mars does or ever did harbor life, and human outposts should be in place. Europa's surface radiation environment is so harsh that humans may never go there, but by 2050 our robots will have begun the exploration of its ocean. We could be in the midst of studying the solar system's second biosphere—and an altogether new biology. We should have the infrastructure in place around numerous worlds to support regular monitoring and exploration. We will have begun our permanent expansion into the solar system, wiser from our experience on Earth, and determined to be better stewards of all the worlds, including our own.

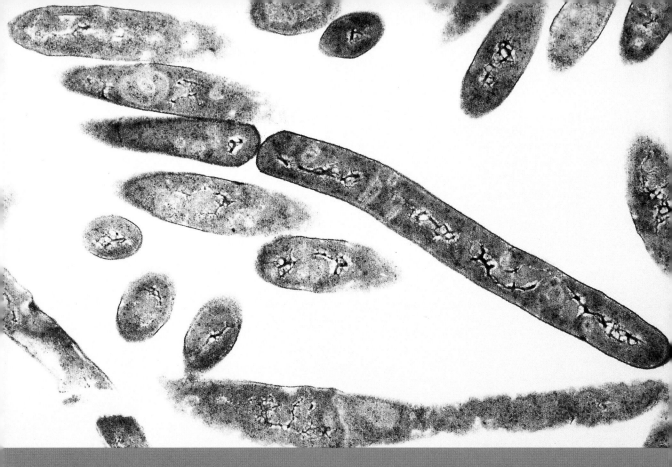

Subsurface Life on the Earth and Other Worlds

We have long been familiar with two domains of life on the Earth: the surface of the land and the body of the oceans. Both domains share the same energy source: namely sunlight, used by green plants and microorganisms in photosynthesis. In this process, sunlight is used to break apart water and carbon dioxide molecules, and the products form the organic molecules and provide the chemical energy that supports the other kinds of life.

This article was adapted from the author's paper "The deep, hot biosphere," published in 1992 in the *Proceedings of the National Academy of Sciences,* vol. 89, pp. 6045–6049. See also his book, *The Deep Hot Biosphere,* published in 1999 by Copernicus Books.

Thomas Gold is Professor Emeritus of Astronomy at Cornell University.

This was the general view of biology until 1977, when another domain of life was discovered. This new domain, the volcanic deep sea vents, proved to have an energy supply that was independent of sunlight and all surface energy sources. The energy for such life is derived from chemical reactions in which hot fluids (liquids and gases), rising continuously from cracks in the ocean floor, combine with substances available in the local rocks and in the ocean water.

How widespread is life based on such internal energy sources? Venting of liquids and gases from the Earth's interior is not limited to cracks in the ocean floor. Deep within the rocks, the fluid-filled fractures and pore-spaces are quite sufficient to accommodate bacterial life, and the rocks themselves contain many of the chemicals that can be nutrients together with the ascending fluids. Deep drilling on land has in fact led to the discovery of bacteria at depths where they were not expected. (The term bacteria refers here to single-celled organisms without a nucleus, including the recently-discovered domain of life called Archaea.)

What are the depths to which bacterial life may have penetrated? Temperature in the Earth increases with depth. Certain bacteria, called hyperthermophiles ("extreme heat lovers"), can live at higher temperatures than any other known organisms. Their survival at 110°C has been verified, and some biologists think the upper limit may be as high as 150°C, providing always that the pressure is sufficient to suppress the boiling of water at these temperatures. Such an upper limit of the temperature occurs between five and ten kilometers below the surface in most areas of the Earth's crust.

Access to such depths would present no problem for bacteria. The normal rate of reproduction of bacterial colonies along cracks and pore spaces would by itself take them down in a few thousand years—a small fraction of the age of these crustal rocks. Fluid circulation through deep cracks would provide much faster transport. Of course, there would be no space in those narrow cracks and tiny pores for larger, multicellular life forms.

How much life could exist within the Earth's crust? Suppose that bacterial life survives down to five kilometers depth. And suppose that the average porosity of the crustal rock is two percent. Then if one percent of that porosity were occupied by bacterial mass, the deep biosphere would be equivalent to a layer of living material one meter thick, if spread out uniformly over the Earth's surface. This would be more than the familiar surface biology that uses sunlight. We do not know how to make a realistic estimate of the amount of subterranean life, but it may be comparable to all the living mass at the surface.

We cannot discuss these possibilities without connecting them with the questions of the origin of life. Photosynthesis is an extremely complex process that must lie some considerable way along the path of evolution. Energy sources that were simpler to tap had to sustain life for all the time from its origin to the perfection of the photosynthetic process. Perhaps these were the same chemical sources that still support life in the ocean vents and in the crust of the Earth.

The rocks that have hydrogen, methane, and other fluids percolating upwards would seem to be the most favorable locations for the first generation of self-replicating organisms. Deep in the rocks the temperature, pressure, and chemical surroundings are constant for geologically long periods of time and, therefore, no rapid response to changing circumstances is needed. No defense is needed against all the

The deep drilling rig in the Siljan impact crater of central Sweden, late 1980s. The bore hole, drilled in Archaean granite, recovered thermophilic bacteria living at a depth below 5,278 meters.

chemical changes caused by solar ultraviolet light or even by the broad spectrum of visible sunlight at the surface.

Microbiologists have found that most of the Archaea—in some ways the most primitive and perhaps ancient bacteria—are hyperthermophiles. This may indicate that life on Earth evolved at such high temperatures in the first place. If it did, and if the Archaea represent the earliest forms of life, evolved at some depth in the rocks, they may have spread laterally at depth, and they may have evolved and progressed upwards to survive at lower temperatures nearer the surface. This may have allowed them to populate all the deep areas that provided suitable conditions to support such life. Of course, now, when the Earth's surface is full of microbial life of all kinds, it may be difficult to unravel the evolution in each of the domains.

When my paper first proposing the existence of a "deep hot biosphere" was published in 1992, a persistent criticism was that microbes brought up in samples from great depths were not native inhabitants but opportunists

introduced from the surface in biologically-contaminated drilling fluids. However, the discovery of heat-resistant bacteria in samples drilled from depths of several kilometers in France, Alaska, Sweden, and elsewhere soon convinced critics that these organisms could not be surface contaminants. Such hyperthermophiles cannot reproduce at the cool temperatures prevailing at the Earth's surface.

It may be that a simple general rule applies: that microbial life exists in all the locations where microbes can survive. That would mean all the locations that have a chemical energy supply and are at a temperature below the maximum one to which microbes can adapt. There would be no such locations on the Earth that have been off limits to life for the long periods of geologic time.

The other planets and satellites in our solar system do not have favorable circumstances for *surface* life. The numerous bodies with solid surfaces all have atmospheric pressures and temperatures that rule out the presence of liquid surface water. This fact long discouraged hopes of finding other life in the solar system.

But with the possibility of *subsurface* life, the outlook is quite different. Many planetary bodies will have temperature and pressure regimes in their interiors that would allow liquid water to exist. Hydrocarbons clearly are plentiful, not only on all the gas giant outer planets, but also on the smaller solid bodies: the satellites, numerous asteroids, Pluto, comets, and meteorites. There is every reason to think that non-biological compounds containing unoxidized carbon were incorporated in all of the planetary bodies at their formation.

The pressures and temperatures at a depth of a few kilometers in the interiors of most of the large solid planetary bodies will not be too different from those in the Earth. These temperatures are not due to the Sun but to internal heat sources—natural radioactivity in the rocks and gravitational compression. The depth at which similar pressures and temperatures will be reached will be deeper in bodies that are smaller than the Earth, but that alone is no handicap for microbial life.

If in fact such life originated at depth in the Earth, there are at least ten other planetary bodies in our solar system that would have had a similar chance for originating microbial life. Is there a possibility of finding life of an independent origin on some of the other planetary bodies? We shall have to see whether microorganisms exist at depths on the Moon, on Mars, in the larger asteroids, and in the satellites of the major planets. Such investigations may become central to that great question of the origin of life, and with that they may become a central subject in future space programs.

It may be some time before the space program can perform deep drilling operations on distant planets. However, there are other options. Deep rifts, such as the Mariner Valley on Mars, expose terrain that was at one time several kilometers below the surface. Samples from the massive landslides in that valley could be returned to Earth and analyzed for chemical evidence that living materials have existed there in the past. Similarly, one may sample deep rifts and young craters on any of the other solid planetary bodies.

Recognizing that even the seemingly most inhospitable bodies may harbor life, we now need to take care to avoid contamination by terrestrial organisms. Manned expeditions, whatever other difficulties they involve, can certainly not be kept sterile, and would therefore spoil such research for all future times. Only very clean unmanned space vehicles

would qualify to bring back meaningful samples of a biology resembling that of the Earth.

The surface life on the Earth, based on photosynthesis for its energy supply, may be just one strange branch of life, an adaptation specific to a planet that happened to have surface conditions that occur only very rarely: a favorable atmosphere, a suitable distance from an illuminating star, a mix of water and rock, etc. The deep, chemically-supplied life, however, may be common in the universe, both in other solar systems and in interstellar planets.

Astronomical considerations make it probable that planetary-sized bodies have formed in many locations from the materials of interstellar molecular clouds, even in the absence of a central star. It is also probable that the formation of solar systems generally causes the escape of as many planets as remain behind in stable orbits. Such planet-sized interstellar objects may be widespread and common in our Galaxy and in others. It is possible that many of them support subsurface life.

Such interstellar planets are so small and dim that we would not have discovered them even if they were so numerous that their combined mass were an appreciable fraction of the total mass of the stars. Perhaps the occasional passage of such "rogue planets" through the outer reaches of our own solar system causes the gravitational disturbances that comets require to bring them into the inner part of the solar system where they become visible. Such interstellar planets could maintain active subsurface microbial biospheres for billions of years, just as seems to be the case on the Earth. The discovery of the subterranean domain of life on Earth vastly increases our estimates of the number of potential sites for life in the universe. ♄

Can Interplanetary Rocks Carry Life?

H. J. Melosh

" [W]e must regard it as probable in the highest degree that there are countless seed-bearing meteoric stones moving about through space. If at the present instant no life existed upon this Earth, one such stone falling upon it might . . . lead to its becoming covered with vegetation. "

—William Thomson (Lord Kelvin), 1871

H. J. Melosh is Professor of Planetary Science at the Lunar and Planetary Lab of the University of Arizona.

This unnamed Martian crater, located on the slopes of the giant volcano Olympus Mons, is twenty-six kilometers in diameter. The impact explosion that made it could have launched meter-sized boulders fast enough to escape from Mars.

In 1883, a tremendous explosion rocked the world as the Indonesian volcano Krakatau collapsed amidst a fountain of incandescent lava and superheated steam. Rakata Island, the remnant stub of the former volcano, was utterly sterilized by the molten rock and hot gases. In the aftermath of this devastating event, biologists watched intently as life slowly returned to Rakata. After six months a single spider, then, after a few years, crabs and trees appeared to repopulate the island. The biologists were fascinated by the conversion of a sterile rock pile to a teeming jungle, but not surprised. Their confident expectation that life would find its way to the devastated island was based on the broad distribution of life on Earth and an understanding of the hardiness of different kinds of seeds and spores under harsh conditions.

Until recently, scientists assumed that the planets in the solar system, unlike islands in the sea, are so isolated that life cannot be transported from one to another by any natural agency. It seemed that life, if we ever found it on other planets, could have arisen there only spontaneously. This possibility excites scientists investigating the origin of life, who would love to study the results of independent origins of life. Unfortunately for them, it now seems that the planets may not be so isolated after all and that the interplanetary transfer of living organisms is a distinct possibility.

This revolutionary view originated with a 1979 proposal that several rare groups of meteorites came from the planet Mars. Although this suggestion was initially greeted with skepticism, evidence for the Martian origin of these meteorites mounted. In 1983 the first meteorites from the Moon were recognized, and most scientists finally admitted the possibility that rocks from one world may occasionally fall onto another. It also became clear that such meteorites, like coconuts washed up on the shore of a deserted island, have the potential of carrying viable microorganisms from Mars to Earth and, by logical extension, from Earth to Mars and elsewhere in the planetary system.

There are now thirteen identified Martian meteorites. They are sometimes called "SNC" meteorites, an acronym from the names of the individual meteorites Shergotty, which fell in India in 1865, Nakhla, which fell in Egypt in 1911, and Chassigny, which fell in France in 1815. These meteorites, named after the post office closest to the recovery site, were recently supplemented by a number of others found in Antarctica, including EETA79001 and ALH84001 (their names designate the collection site, year of collection, and catalog number).

The remarkable thing about the Martian meteorites is that they are so unlike other meteorites and so similar to Earth rocks. The Shergottites (which include Shergotty itself, Zagami, which fell in Nigeria in 1962 and EETA79001, found in Antarctica in 1979) would be nearly indistinguishable from a terrestrial basalt if they had not obviously fallen from the sky. All other Martian meteorites similarly resemble Earth rocks that have crystallized from magmas or, in the case of ALH84001, have been altered by the deposition of carbonates from hot underground water.

The main difference between the Martian meteorites and Earth rocks is in the physical state of their minerals. Most minerals in the meteorites are thoroughly shattered on a microscopic scale, reflecting the force of the explosion that ejected them from Mars. In EETA79001, the blow was so strong that it collapsed pre-existing small cavities and melted some of the surrounding rock. This melt trapped the atmospheric gases in the cavity and then froze solid, preserving the gases during the long journey from Mars to the Earth. Indeed, it was the discovery of these dissolved gases and the recognition that they are identical to the Martian atmosphere, as measured by the Viking lander on the Martian surface in 1976, which tipped the balance of scientific opinion into accepting a Martian origin for these meteorites.

Once a Martian origin was established, another problem immediately arose: How did these rocks get thrown off the surface of a planet? One obvious suggestion was volcanic eruption. After all, most of these meteorites are essentially volcanic rocks. We know that Mars has massive volcanoes and that volcanoes on Earth can toss rocks out to great distances. Could some mega-eruption on Mars have blown a bunch of rocks out into space? The answer is no.

For rocks to be spewed from volcanoes at speeds high enough to achieve escape velocity from a planet, the volcanic gases driving the rocks must expand at least as fast as the rocks themselves are moving. However, the maximum expansion velocity of a gas is related to its temperature. Although a terrestrial volcano, with expanding water vapor at temperatures up to 1,200°C, can eject volcanic rocks at nearly 1 km/sec (2,200 mph), a Martian volcano would need water vapor at nearly 40,000°C to accelerate rocks to the Martian escape velocity of 5 km/sec. No geophysicist could accept such high temperatures for the Martian mantle, so volcanic ejection is not a plausible mechanism.

The only other known natural process that can accelerate surface materials to such high speeds is meteoritic impact. When a large, high speed meteorite strikes the surface of a planet,

the energy of its motion is instantly converted to heat and motion in both the projectile and target materials. The resulting explosion opens a crater in the surface, expelling target material at high speed from the growing cavity. Most of the fast-moving material is either melted or vaporized. Even if it escaped the planet, it would not be recognized as a rock. However, a small amount of material originally on or close to the surface near the meteorite's impact point is protected from high pressure. Because it has little or no rock lying above it, this surface material can never be compressed during the explosion to pressures high enough to totally destroy it. Instead, like a watermelon seed squeezed between two fingers, it slips out and shoots upward into space at high speed. Deeper-lying materials are squeezed harder before being shot upward, until at large enough depths the free surface is too far away to prevent destruction of the rocks.

By this process, called "spallation," small quantities of rock near the surface in the zone around a meteorite impact are thrown out at high speed with only slight impact shattering. Some of this material can exceed the escape velocity of the Moon, Mars, or even the Earth. Even the Earth's atmosphere does not prevent large impacts from ejecting material into space. Along with the spalled surface rocks, a powerful ball of gas from the vaporized projectile and target expands out of the crater, pushing aside the surrounding atmosphere and giving the speeding rocks free access to space.

Unlikely as this scenario may seem, we have evidence supporting it. The Ries impact crater, twenty-two kilometers (fourteen miles) in diameter, was formed in what is now southern Germany by an impact some 15 million years ago. It ejected meter-sized blocks of limestone that were recovered in Switzerland, 200 kilometers from the impact site. As predicted by the theory of spallation, these limestone blocks come from the uppermost geologic unit at the impact site. Although these blocks did not leave the Earth (or we would never have found them), they must have traveled in ballistic trajectories above the atmosphere for most of their flight, and thus it seems certain that spalled surface rocks can make it through the Earth's atmosphere. Furthermore, dark patches of ejected debris extend far enough from large impact craters on Venus to suggest that such material can breach even the extremely dense Venusian atmosphere.

The spallation mechanism for launching meteorites has now been studied experimentally in laboratory impacts and numerically by the same computer codes used to model nuclear explosions. The lunar meteorites were all clearly lying on the surface of the Moon before launch, as judged by the distribution in them of elements produced by exposure to cosmic rays. The Martian rocks came from deeper strata, but still must have been close to the surface at the time of launch. All this is consistent with the spallation process and with the idea that the Martian and lunar meteorites were launched by the impact of larger meteorites. These impacting meteorites must have been at least a few hundred meters in diameter in order to launch the meter-sized Martian meteorites found on Earth.

A fascinating implication of this launch mechanism is that a small but important fraction of the high-speed debris is shot out into space with almost no damage at all. Although most of the Martian meteorites show signs of the blow that launched them, others do not. Nakhla, in fact, contains traces of fragile hydrated carbonate minerals formed by weathering on the Martian surface. These minerals would be destroyed if they had ever been raised to temperatures over about 200°C. If these minerals could survive launch, it is not

farfetched to suppose that living organisms, such as bacteria or bacterial spores, could survive as well.

If living organisms can survive ejection from the surface of a planet, then the idea of strict biological separation of the planets must be questioned. Of course, larger organisms, such as plants and insects, stand no chance of hitching a ride on one of these rocky space probes. The acceleration upon launch is tens of thousands of times larger than the Earth's surface gravity (g), so that differential forces would squash anything larger than a few microns in size. However, microbes are regularly accelerated to hundreds of g in laboratory ultracentrifuges without apparent harm. Recent experiments in launching spores and microbes from a cannon show that such a large acceleration is not fatal to the right species.

Just surviving the launch is not enough, however. Even if bacteria and bacterial spores present in, say, a crack or pore of a near-surface rock were launched by the nearby impact of a large meteorite, they would have to endure the vacuum of space and harsh solar ultraviolet and cosmic radiation for perhaps millions of years before the rock encountered another planet. However, it seems that Earth life is actually up to this task.

Vacuum itself presents few problems for bacteria. Indeed, the preferred method of preserving bacterial cultures in laboratories is simply to subject them to a vacuum. The Apollo 12 astronauts recovered terrestrial bacteria preserved in this way from the unmanned Surveyor 3 lander. Even after exposure for three years on the lunar surface, these bacteria were successfully cultured in the laboratory. Other experiments on Skylab and NASA's Long Duration Exposure Facility show that the principal threat to organisms in space is ultraviolet (UV) radiation, not vacuum.

However, UV is easily stopped by just a few microns of dust. Organisms hiding inside a rock are safe from this kind of radiation.

Cosmic radiation is a more serious threat. Low energy solar cosmic rays do not penetrate deeply into a rock, but galactic cosmic rays smash their way meters deep into rock and could limit the survival of most bacteria by degrading their genetic material. However, some bacteria are highly resistant to radiation damage. *Deinococcus radiodurans* (named for its radiation resistance) has been found living in the cores of nuclear reactors. Even the common soil bacterium *Bacillus subtilis* forms radiation-resistant spores. Such resistance to radiation is apparently associated with resistance to drying, so that many desert soil bacteria seem almost pre-adapted for a journey through space. Without this resistance, bacteria would be limited to traveling in rocks many meters in diameter, large enough to shield them from cosmic radiation. Some rare impact events do eject rocks of this size into space, so perhaps radiation resistance is not an absolute requirement for interplanetary travel.

Finally, the dormant bacteria must survive long enough to reach a new planet. Few of the rocks ejected by an impact have enough speed to take them directly to another planet. Instead, planetary meteorites will first revolve around the Sun in orbits similar to that of their mother planet. In time, the planet may recapture them. But more often they are flung into new orbits by the planet's gravity. This scenario may repeat itself many times before the orbit is changed so much that the meteorite crosses the orbit of another planet, where this cosmic dance continues with a new partner. Eventually, most such meteorites collide with another planet or the Sun. Some may be flung out of the solar system entirely.

However, all this interplanetary wandering takes millions of years, and for dormant bacteria or spores inside a rock, the DNA degradation clock is ticking. We do not yet know how long dormant organisms may survive. There are some indications that the survival time may be as long as millions of years, in which case a large percentage of the debris ejected from a life-bearing planet will contain viable organisms. Other experts believe the survival time is much shorter, centuries or millennia, in which case only a tiny fraction of the material from one planet could infect another.

However, the number of individual fragments ejected by a single impact is enormous, and the number of bacteria and spores likely to be present in surface rocks is immense. A recent analysis of this problem by a team of American and European experts has determined that viable organisms from the Earth have almost certainly reached Mars at some time in the history of the solar system. And if life first developed on Mars, it has likely reached the Earth by now. The proportion of meteorites known to be from Mars suggests that about half a ton of Mars rocks are now falling on Earth each year. If life ever existed on Mars, it has already had the opportunity to colonize the Earth.

The final act in this interplanetary drama is the rock's fall onto its new planetary home. Planets with atmospheres, like the Earth and Mars, have an advantage since the entering rock decelerates relatively gently and may even break up high above the surface, exposing its previously concealed interior. Once their cosmic velocity is dissipated by flight through the atmosphere, the fragments of the broken rock fall relatively gently to the surface. This atmospheric passage is so short that the interiors of most fragments are not even warmed by the brief heating during entry. Indeed, most people fortunate enough to encounter a freshly fallen meteorite are surprised to discover that it retains the frigid cold of space. Any dormant microorganisms now have an opportunity to awaken and colonize their new environment.

Mars is now inhospitable. Its atmosphere is unbreathable and so thin and cold that liquid water cannot exist on the surface. Deadly solar ultraviolet radiation bathes the surface, and the soil is full of highly-oxidized chemicals that would quickly degrade organic material. Microorganisms arriving in a meteorite from Earth today would quickly perish upon exposure. However, Mars was not always this way. Images from Viking and Mars Surveyor show a planet covered with traces of past surface water. The scars of ancient floods, river channels, shorelines of dried-up lakes and perhaps seas, are clearly visible. At one time Mars must have been much warmer and wetter. Life from Earth may well have found a haven on such a planet.

Can such a story be proved? The only way to learn if life really traveled between Earth and Mars is to look for life on Mars. That will not be easy, because nothing can live on the Martian surface now. We will have to dig below the surface, collect and examine rocks from ancient lakes and seas, seek hot springs, and drill for subsurface water. Ideally, we will find living organisms, the DNA of which can be compared to that of organisms on Earth (assuming they have DNA). Perhaps we will find only fossil remains of an extinct biota, which may make determination of its origin more difficult. In any case, the answers to these questions lie on Mars and other possible havens for life in the solar system, such as the giant moons Europa or Titan. To answer these deep questions about life, we must go there and look.

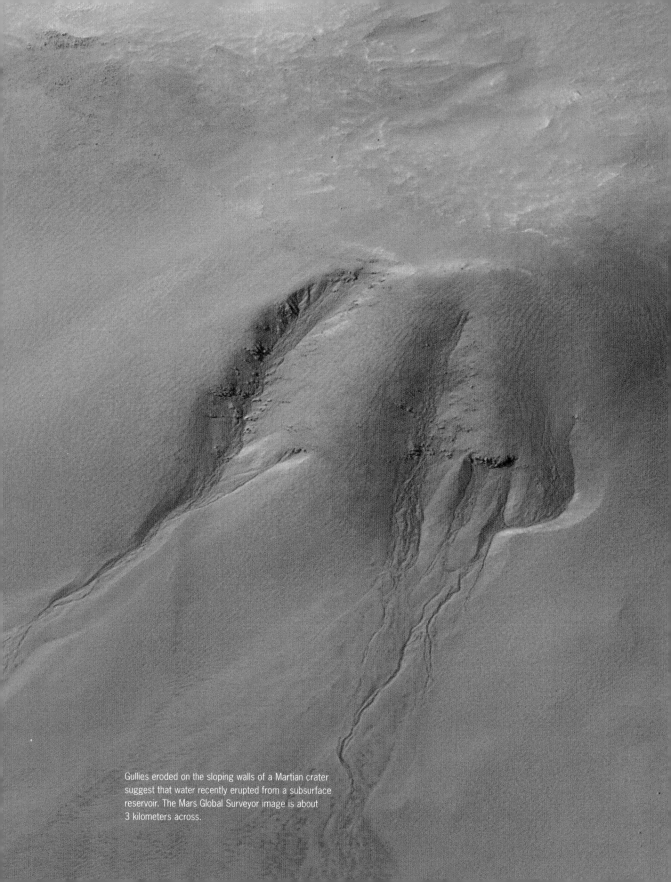

Gullies eroded on the sloping walls of a Martian crater suggest that water recently erupted from a subsurface reservoir. The Mars Global Surveyor image is about 3 kilometers across.

Martian meteorite ALH84001, recovered in Antarctica. Some scientists have suggested that physical and chemical features in this meteorite provide evidence for microscopic fossil life on Mars. That interpretation remains controversial.

Fossil Microbes from Mars?

In 1996, a team of scientists led by David McKay of NASA's Johnson Space Flight Center announced that they had discovered evidence for microscopic fossil life in a meteorite from Mars. From the start, the evidence was both fascinating and controversial, and to this day it remains so.

The meteorite in question had escaped from Mars 16 million years ago when an asteroid or comet collided with the planet and blasted out a crater. The 2-kilogram fragment of Martian rock then moved in an elliptical orbit around the Sun until it was swept up by the Earth about 13,000 years ago. It landed in glacial Antarctica, where it remained until 1984, when a meteorite-hunting party picked it up it in the Allan Hills. The specimen was designated ALH84001. At first, no one suspected that it came from Mars. About ten years later, scientists examined

ALH84001 more closely and found that it was not an ordinary meteorite, but one of the so-called SNC meteorites, which come from Mars. Meteorites of this class all contain traces of gas having a composition identical to the Martian atmosphere. Each of the dozen other SNC Martian meteorites then known had crystallized within the last 1.3 billion years, after Mars had become a frozen desert. But ALH84001 was over 4 billion years old, and had presumably existed at a time when liquid water was common on the surface of Mars. Liquid water is essential to life as we know it. For that reason ALH84001 attracted the attention of McKay and his team, who thought that the rock might preserve microscopic and chemical evidence of ancient life on Mars.

To avoid the possible terrestrial contaminants picked up by the meteorite in Antarctica, the team obtained their samples from the solid interior volume of the rock. They found that cracks within the meteorite contain orange-tinted carbonate globules, which resemble limestone cave deposits. This sort of material can form only in the presence of liquid water. McKay and his colleagues found three kinds of evidence that they interpreted in terms of ancient microbial life on Mars:

1   *The globules contained traces of complex organic compounds called polycyclic aromatic hydrocarbons (PAHs), which might be the decay products of microbes.*

2   *The globules contained microscopic grains of magnetite (a magnetic iron oxide) and of iron sulfide, two compounds rarely found together in the presence of carbonates, unless produced by bacterial metabolism.*

3   *The carbonate globules, when examined with an electron microscope, were found to be covered in places with large numbers of worm-like forms that resemble fossilized bacteria.*

McKay and his colleagues conceded from the start that any one of these lines of evidence could be interpreted without recourse to biology. However they judged that the presence of all three in association with the carbonate globules made a persuasive case for ancient life on Mars. Other scientists at once began to subject the evidence to intense critical scrutiny, which is an expected and essential part of the scientific process.

Some geochemists found evidence that the carbonate globules were formed at temperatures up to 300° C, too hot for the survival of any known microbial life on Earth. But others con-cluded that the globules may have formed at temperatures below 100° C. This now seems to be the case.

What about the PAHs? Chemical theory and experiments show that organic (carbon-based) molecules are formed non-biologically within giant interstellar clouds of gas and dust. The solar system was born in such an environment. Organic molecules produce PAHs when heated, so these materials would have been present on the Earth and on Mars from the beginning. Ordinary soot contains PAHs. McKay and his colleagues soon conceded that the organic carbon evidence for fossil life on Mars was weak.

The evidence from the magnetite and iron sulfide grains is more substantial. In size and shape, most of the magnetite grains closely resemble those produced by terrestrial bacteria. No one has yet demonstrated a non-biological formation mechanism for such grains in association with iron sulfide. The evidence so far suggests that such materials require a biological origin.

What about the objects said to resemble fossil bacteria? Microscopists contend that shape alone is often misleading, because non-biological processes can produce objects that superficially resemble bacteria. The size of the supposed Martian fossils is also a contentious issue. Many of the worm-like objects in ALH84001 are just a few tens of nanometers across, or about a tenth the size of the smallest-known bacteria on Earth. But the minimum amount of molecular "equipment" needed to keep a bacterium alive (including the DNA and ribosomes to translate the genetic code into proteins) would require a volume equal to a 200-nanometer sphere. In other words, these so-called Martian fossils are just too small to have ever been alive.

But just a moment. How can we be sure that any hypothetical Martian life would use the same kind of biochemistry as terrestrial bacteria? Does all life have to be based on molecules as large and complex as DNA? Some scientists have reported finding so-called "nanobacteria" in a wide range of environments. These mysterious objects are as small as the alleged Martian microbes, and are conceivably living organisms, or fragments of organisms. Before the evolution of DNA, ribosomes, and complex proteins on Earth, simpler ancestral life forms must have existed. Those primitive organisms would have lost out in competition by the far more complex bacteria that later evolved on Earth, but their fossilized remains might still be found. And perhaps such things have been found in the Martian meteorite.

For several years, the general current of scientific opinion was running against the biological interpretation of the evidence from ALH84001. But recent studies of the magnetite grains provide increased support to a biological origin. Scientists continue to study ALH84001 and the other Martian meteorites. And they are planning a mission to bring samples back from Mars. While the evidence from ALH84001 remains controversial, it has without question stimulated a major new effort in the search for life, ancient or extant, on Mars.

The possibility that life exists today on Mars received a boost with the publication in June 2000 of orbiter images showing geologically fresh erosion channels on the slopes of Martian craters (see page 176). The evidence suggests that there may be liquid water in geological formations not far below the surface of Mars. Perhaps this water breaks out at the surface intermittently and excavates channels before it rapidly boils into the thin Martian air. And where there is subsurface liquid water, there may be life. Stay tuned. 🪐

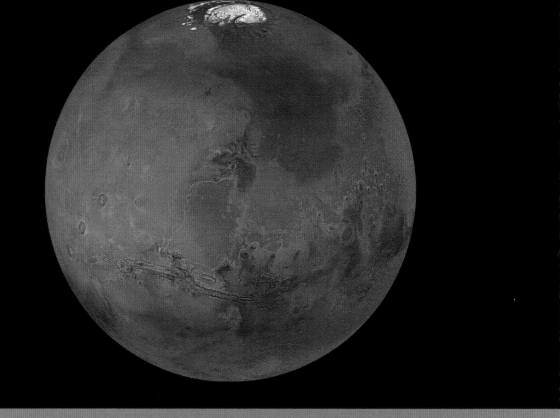

Bringing Life to Mars

Christopher P. McKay

Four billion years ago Mars was a warm and wet planet, possibly teeming with life. Spacecraft orbiting Mars have returned images of canyons and flood valleys—features that suggest that liquid water once flowed on the planet's surface. Today, however, Mars is a cold, dry, desertlike world with a thin atmosphere. In the absence of liquid water—the quintessential ingredient for life— no known organism could survive on the Red Planet.

More than twenty years ago the Mariner and Viking missions failed to find evidence that life exists on Mars's surface, although

Christopher P. McKay is a Research Scientist with the Space Science Division of the NASA Ames Research Center.

This image of Mars, with the north polar ice cap at top and the great equatorial canyon system (Mariner Valley) below center, was compiled from about 1000 images taken by the Viking orbiter spacecraft.

all the chemical elements needed for life were present. That result inspired biologists Maurice Averner and Robert D. MacElroy of the National Aeronautics and Space Administration Ames Research Center to consider seriously whether Mars's environment could be made hospitable to colonization by Earth-based life-forms. Since then, several scientists, using climate models and ecological theory, have concluded that the answer is probably yes: With today's technology, we could transform the climate on the planet Mars, making it suitable once more for life. Such an experiment would allow us to examine, on a grand scale, how biospheres grow and evolve. And it would give us the opportunity to spread and study life beyond Earth.

Why Mars?

Many of the key physical properties of Mars are remarkably similar to those of Earth (see Table A on page 186). On both planets the length of day is about 24 hours—an important consideration for plants that have adapted to photosynthesize when the sun shines. Mars also experiences seasons, as the planet's axis is tilted to a similar degree as Earth's. Because Mars is farther from the sun, a Martian year is almost twice the length of an Earth year, but plants should be able to adapt to such a difference. One unalterable difference between Earth and Mars is gravity: Martian gravity is about one third that of Earth's. How life would adapt to reduced gravity is unknown. It is likely, however, that microbes and plants would adjust easily to Martian gravity, and some animals might cope just as well.

Other planets and moons in our solar system also might be considered potential sites for life including Venus, Titan, and Europa. Each of these bodies, however, possesses some basic physical parameter that is inconsistent with habitability. Titan and Europa—satellites of Saturn and Jupiter, respectively—are too far from the sun. Venus is too close, and its extremely dense atmosphere makes the planet much too hot for life. Furthermore, the planet rotates so slowly that its day is equal to about four months on Earth, which might make life difficult for plants. The technology needed to alter these physical parameters is well beyond the current scope of human capability.

Mars is currently too cold, too dry, and its carbon dioxide atmosphere too thin to support life. But these parameters are interrelated, and all three can be altered by a combination of human intervention and biological changes. The key is carbon dioxide. If we were to envelop Mars in a thicker carbon dioxide atmosphere, with a surface pressure one to two times that of air at sea level on Earth, the planet would naturally warm above the freezing point of water. Adding a bit of nitrogen to the atmosphere would help satisfy the metabolic needs of plants and microbes. And the small amount of oxygen that would be produced from the photochemical degradation of carbon dioxide could create a rudimentary but effective ozone shield for the rejuvenated planet. This carbon dioxide atmosphere would support plant and microbial life but would not contain enough oxygen for animals.

Although humans would need to carry a supply of breathable air with them, a carbon dioxide Mars would still be a much kinder, gentler place than today's Mars. The higher temperatures and atmospheric pressure would make bulky space suits and pressure domes unnecessary. And the natural growth of plants would allow the cultivation of farms and forests on Mars's surface, thus providing for human colonists or visitors.

To make Mars suitable for animals and humans, its atmosphere would have to be made more similar to Earth's, which is composed primarily of nitrogen, with oxygen levels close to twenty percent and carbon dioxide levels less than one percent. The process of generating such an Earth-like, oxygen-rich environment—also called terraforming—would be much more difficult than simply thickening Mars' atmosphere. But to make Mars habitable, generating a carbon dioxide atmosphere—a process that biologist Robert Haynes of York University has dubbed ecopoiesis—would be the logical first step.

Does Mars possess the essential volatiles—carbon dioxide, nitrogen, and water—needed to create a habitable environment? Ferrying these raw materials from Earth would be impractical. For example, the amount of nitrogen needed to create a breathable atmosphere on Mars is more than a million billion tons. The space shuttle can carry only about twenty-five tons into low-Earth orbit. Thus, if Mars does not have the necessary amount of nitrogen, it is not within near-term capabilities of humans to bring it there.

Unfortunately, we do not yet know how much of each of these key ingredients Mars has hidden below its surface. We do know that the thin Martian atmosphere currently contains only small amounts of carbon dioxide, nitrogen, and water vapor. But at one time Mars must have had a much thicker atmosphere. Researchers have used a variety of methods to estimate how much carbon dioxide, nitrogen, and water would have been present in the early Martian atmosphere. These methods—which include measuring the ratio of nitrogen isotopes and estimating the volume of water needed to etch the Martian flood channels—yield widely different estimates of the amount of volatiles once present on the planet.

Fortunately, the range of estimates overlaps the

amounts of volatiles needed to produce a breathable atmosphere and a substantial ocean (see Table B on page 186). It is possible that some of these volatiles have left the planet permanently, flowing out into space because of Mars's low gravity. If, however, Mars once had enough of the volatiles needed to make a biosphere, it probably still has them locked up in the subsurface. Water could be frozen as ground ice, and nitrogen could be contained in nitrates in the Martian soil. Carbon dioxide could be frozen in Mars' polar caps as well as in the soil.

Turning up the Heat

If Mars does not have the essential ingredients, the first step in transforming the environment is to warm the planet. Heating the Martian surface would release the carbon dioxide, nitrogen and water vapor in the atmosphere. The energy needed for such massive heating would have to come from the sun. Compared with sunlight, human energy sources are small. For example, sunlight delivers more energy to Mars in thirty minutes than the energy that would be released by the explosion of all the nuclear warheads of the U.S. and Russia. So trapping the energy from sunlight and using it to warm the planet is really the only practical option for generating a life-friendly Mars.

Through the years, scientists have proposed and considered several methods of using sunlight to heat Mars. Some researchers suggested spreading dark soot on the polar caps to help them absorb more sunlight and melt their stores of frozen carbon dioxide. Other researchers proposed putting large mirrors in orbit around Mars to reflect sunlight on the polar regions. But the technologies needed for these methods have never been demonstrated. The space mirror, for example, would have to be the size of the state of Texas to increase the amount of sunlight hitting Mars by just two percent.

Perhaps the most practical approach to warming Mars would involve using "super-greenhouse" gases to trap solar energy on the planet. This method was first suggested by British atmospheric scientist James Lovelock, who is best known for the Gaia hypothesis that the presence of life maintains the habitability of Earth. Lovelock's idea for heating Mars involved pumping gases such as methane, nitrous oxide, ammonia, and perfluorocarbons (PFCs) into the Martian atmosphere. These super-greenhouse gases can trap solar energy with thousands of times the efficiency of carbon dioxide, the most abundant greenhouse gas on Mars and Earth. Even small amounts of the super-greenhouse gases can warm a planet; in fact, many scientists believe that the production of these gases is contributing to global warming here on Earth.

Computer calculations performed by myself, Owen B. Toon and James F. Kasting suggest that if Mars's atmosphere contained just a few parts per million of the super-greenhouse gases, the average temperature at the planet's surface would rise from –60 to –40 degrees Celsius (–76 to –40 degrees Fahrenheit). This warming could be enough to trigger the release of carbon dioxide from the polar caps and soil into the atmosphere. Carbon dioxide would then augment the greenhouse effect even further, driving the release of more carbon dioxide and water vapor into the atmosphere. Such positive feedback would be sufficient to create a thick, warm atmosphere—the carbon dioxide Mars.

Where would the greenhouse gases come from? Although PFCs at a concentration of a few parts per million would do the job, the mass of material needed to warm Mars would be much too large to import from Earth. Instead the greenhouse gases would have to be produced locally, on Mars—chemically at first and eventually

biologically, with the help of microorganisms. The gases must be easily synthesized from elements likely to be abundant on Mars and must persist in the Martian atmosphere for a relatively long time. PFCs such as CF_4 and C_2F_6, and other compounds such as SF_6, would be good choices because they absorb thermal radiation efficiently and would have long lifetimes in the Martian atmosphere, on the order of hundreds of years. Furthermore, the elements making up these compounds—carbon, fluorine, and sulfur—are all abundant on Mars.

To generate enough greenhouse gases, we would need to distribute hundreds of small PFC factories across the Martian surface. Powered by solar energy, each of these Volkswagen-size machines would harvest the desired elements from Martian soil, generate PFCs and pump these gases into the atmosphere.

The Matter of Time
How long would it take to generate a thick carbon dioxide atmosphere? The atmospheric PFCs would have to heat the planet enough to melt the carbon dioxide and water frozen in the polar caps and to evaporate nitrogen from the soil. But how much energy is needed to raise the temperature of Mars? According to our calculations, defrosting Mars would require an energy input of five megajoules per square centimeter of planetary surface. This amount of energy is equivalent to about ten years worth of Martian sunlight.

Trapping this energy would vaporize the frozen carbon dioxide, generating enough gas to create a thick atmosphere. If enough carbon dioxide were generated to provide a pressure twice that of Earth's atmosphere, the average Martian surface temperature would rise to an Earth-like 15 degrees Celsius. At this stage, the bulk of the planet's water is likely to still be frozen deep underground, where temperatures would remain much lower. Melting the reservoirs of subsurface ice would require an additional twenty-five megajoules per square centimeter of surface, equivalent to fifty years of Martian sunlight.

Thus, if every photon of sunlight reaching Mars were captured with one hundred percent efficiency, the planet could be warmed in a decade and fully thawed in sixty years. Of course, in reality no process is one hundred percent efficient. If greenhouse gases can trap sunlight with an efficiency of ten percent, using PFCs could generate a thick carbon dioxide atmosphere in about 100 years and lead to a water-rich planet in about 600 years. These numbers are encouraging. If the answers had turned out to be millions of years, we would have to abandon our plans to turn Mars into a second home for life.

For even quicker results, the greenhouse gas effect could be amplified by coupling it with other methods, such as the deployment of huge orbital mirrors or the spreading of dark material on the planet's surface, according to calculations by Robert Zubrin. But changing Mars slowly makes sense for a number of reasons. Transforming the climate of Mars over decades and centuries—as with greenhouse gases—would be financially feasible. NASA's Mars program could easily absorb the cost of shipping half a dozen PFC factories to the planet every year. Furthermore, working with longer timescales would also allow life on Mars to adapt and evolve and interact with the environment—as has been the case on Earth for billions of years. Finally, slowing the process of environmental evolution gives us ample opportunity to study the coupled biological and physical changes as they occur. Learning how biospheres are built is part of the scientific return for the investment in bringing Mars to life.

Plants and bacteria can thrive on this warm, wet, carbon dioxide-rich Mars. But producing an oxygen-rich atmosphere capable of supporting animals—and humans—is much more difficult. Thermodynamic calculations indicate that conversion of the carbon dioxide in Mars's thick atmosphere to oxygen would require about eighty megajoules of energy per square centimeter, or about 170 years of Martian sunlight. And the only mechanism that could transform the entire atmosphere is a planetwide biological process: the photosynthesis done by plants, which take in carbon dioxide and expel oxygen.

On Earth, the efficiency with which plants produce oxygen from sunlight is only a hundredth of one percent. With this efficiency, converting Martian carbon dioxide to oxygen would take more than a million years. Although this may sound like a long time, keep in mind that the same process on Earth took over two billion years. Of course, as plants consume the atmospheric carbon dioxide, the greenhouse effect would lessen, and Mars would once again become cold. To keep the surface temperatures warm with an atmosphere that contained mostly nitrogen and oxygen and only one percent carbon dioxide, the concentrations of supergreenhouse gases would have to be maintained at a few parts per million. Such quantities of greenhouse gases would be harmless to living things.

Future Martians

If Mars is currently a planet bereft of life, the Martians of the future would have to be imported from Earth. The dry valleys of Antarctica—the coldest, driest and most Mars-like place on Earth—harbor some ideal candidates for the first generation of Martians. High in the mountains, where the air temperature rarely rises above freezing, E. Imre Friedmann of Florida State University has found lichens and algae that live a few millimeters below the surface of porous sandstone rocks. When sunlight warms these rocks, enough snow melts into the sandstone to provide the moisture the microbes need to survive. Similar microorganisms that can grow without oxygen might be able to survive in their little "rock greenhouses" even in the early stages of Mars's transformation, when the planet would still be very cold.

As Mars warms, different types of plants could be introduced. James M. Graham of the University of Wisconsin likens the gradual greening of Mars to hiking down a mountainside on Earth. As one descends to lower elevations, the temperature rises and the scenery grows more lush. On Mars, the bare rock would give way to the hardy plants that thrive on Earth's tundra, and eventually the Martian landscape would blossom into the equivalent of an alpine meadow or a pine forest. The plants would generate oxygen, and eventually insects, worms and other simple animals that can tolerate high concentrations of carbon dioxide and low levels of oxygen could roam the planet.

Introducing life to Mars would be of great scientific merit and could well be relevant to understanding how to sustain the biosphere of Earth. But would such a program be desirable? What are the ethical considerations surrounding such a drastic alteration of another planet's environment?

First we must assume that Mars is currently lifeless—an assumption that must be certified to a high level of confidence before we transfer life from Earth. If Mars did harbor living organisms beneath its surface, we might consider altering the environment to allow that native life to emerge and spread across the planet. If, however, Mars has no life and we believe that life in itself has intrinsic worth, then a Mars replete with life could be considered of more value than today's Mars, beautiful but lifeless.

On Earth, environmental change almost always produces some negative effects. Would this also happen on Mars? Although we can monitor Mars as it evolves, we will not really be able to control or predict the paths that the biota and environment will follow. The Earth's biosphere is so complex that unintended changes that adversely affect some life-forms are to be expected. But on Mars, in all likelihood, no life-forms currently exist. Thus, any biological expansion would be considered an improvement. If spreading life is the objective, making Mars habitable might allow humans to make a purely positive contribution for once.

A Futile Effort?

Billions of years ago Mars had a thick carbon dioxide atmosphere and temperatures warm enough for liquid water. Why then did it become uninhabitable? And if we restored a more hospitable Martian climate, would the planet once again revert to its current barren state?

The answers lie in carbon recycling. Atmospheric carbon dioxide reacts with liquid water to form carbonic acid. This acid weathers rocks, ultimately producing calcium carbonate. As this mineral accumulates in the oceans and lake basins, carbon is effectively removed from the atmosphere.

On Earth, carbonates are recycled by plate tectonics. Subduction of oceanic plates under continental plates carries the sediments deep underground, where temperatures greater than 1,000 degrees Celsius convert the carbonates back to carbon dioxide. Mars, however, is a one-plate planet with a single thick crust.

Because Mars had no plate tectonics to recycle carbonates, it gradually lost its atmospheric carbon dioxide. As the atmospheric pressure dropped, the planet's surface chilled, and its liquid water froze.

If Mars were warmed and its thick carbon dioxide atmosphere restored by human effort, it is very likely that carbonate formation would again deplete the atmosphere. After a few hundred million years, Mars would once again lose its capacity to support life.

But 100 million years is a long time. In fact, Earth might not be habitable for much longer than that. As the sun continues to brighten, Earth will succumb to a runaway greenhouse effect. The oceans will evaporate, creating Venus-like conditions unsuitable for life. So our second planetary home might last almost as long as our first. —C.P.M.

Mars

Earth

Venus

Table A

Comparing	Earth	Mars	Venus
Gravity (g's)	1	0.38	0.91
Length of day	24 hours	24 hours 37 minutes	117 days
Length of year	365 days	687 days	225 days
Axis tilt (degrees)	23.5	25.2	2.6
Average sunlight reaching the planet (watts per square meter)	345	147	655
Average surface temperature (degrees Celsius)	15	−60	460
Surface pressure (atmospheres)	1	0.008	95
Most abundant gases in atmosphere	Nitrogen, oxygen	Carbon dioxide	Carbon dioxide

Table B

The Essential Ingredients for Life on Mars			
	Carbon dioxide surface pressure (atmospheres)	Nitrogen surface pressure (atmospheres)	Water ocean depth* (meters)
Amount needed for plant and microbe habitability	2	0.01	500
Amount needed for breathable atmosphere	0.2	0.3	500
Amount in the present Mars atmosphere	0.01	0.00027	0.000001
Range of estimates for amount on Mars at planet's formation	0.1–20	0.002–0.3	6–1,000

* Amount of water is measured in terms of the depth of an ocean covering the entire surface of Mars.

Carl Sagan and the Quest for Life in the Universe: Profile

Carl Sagan (1934–1996), American planetary astronomer, exobiologist, popular educator, and advocate for science.

" Ask courageous questions. Do not be satisfied with superficial answers. Be open to wonder and at the same time subject all claims to knowledge, without exception, to critical scrutiny. Be aware of human fallibility. Cherish your species and your planet. "

—Carl Sagan

He was a leading planetary astronomer, a pioneer in the search for extraterrestrial biology, a spellbinding teacher, and the most effective public advocate for the values of science the world has ever seen. To hundreds of millions of people, Sagan communicated his passion for the universe of science. "When you're in love," he said, "you want to tell the world."

Carl Sagan (1934–1996) grew up in a working-class family in Brooklyn. At age seven, he went to the public library to find out what the stars are. The answer—that the stars are suns, only very far away, and the Sun is a star, but close up—opened boundless vistas in his young mind. He understood that if those countless stars are suns, they might have their own planets. The universe could be teeming with life. The idea was delectable.

Sagan also learned about a powerful method, called science, that could help him explore such ideas. He knew then what he wanted to do with his life, and he prepared himself well. He went to the University of Chicago, where he studied biology and physics, and earned his Ph.D. in astrophysics in 1960. His mentors were the geneticists Hermann Muller and Joshua Lederberg, the geochemist Harold Urey, and the planetary astronomer Gerard Kuiper. Three of them were Nobel Laureates.

As part of his wide-ranging Ph.D. thesis, Sagan solved an outstanding puzzle in solar system astronomy: Why is Venus such a strong source of microwave radiation? At the time, it was widely assumed that Venus had a warm and wet climate, a plausible enough environment for life. But Sagan calculated that the dense carbon dioxide atmosphere of Venus sustains an extreme greenhouse effect, which keeps the surface hot enough to melt lead and to emit the observed microwaves. Many years later the Pioneer Venus spacecraft verified this explanation. Surely nothing could live near the surface of Venus.

After teaching genetics at Stanford University's School of Medicine, Sagan joined the astronomy faculty at Harvard University, where he gave a series of popular lectures called "Planets as Places." This was a radical idea at the time. Few scientists had thought seriously about the geology and climates of other worlds. Few if any had recognized that the study of other planets could provide vital clues for understanding the Earth. By 1963, Sagan was already concerned that increasing carbon dioxide in the Earth's atmosphere would lead to serious global warming.

These were the years when the spacecraft exploration of the solar system was just beginning. Sagan became a familiar figure at NASA's Jet Propulsion Laboratory in Pasadena, where he was a principal investigator in every American spacecraft mission to the planets, including the Mariner flybys of Venus and Mars, the Viking orbiters and landers sent to Mars, and the Pioneer and Voyager missions to explore the outer solar system.

In 1967, Sagan and James Pollack, his first graduate student, solved another major mystery of the solar system: What causes the seasonal "wave of darkening" observed on Mars? The most popular view ascribed the phenomenon to seasonal changes of vegetation on the planet. But Sagan and Pollack proposed instead that seasonal winds alternately deposit light-colored Martian dust on darker highland rock and then remove it again. This explanation was later verified by the Viking spacecraft in orbit around Mars.

In 1968, Sagan joined the astronomy department at Cornell University. There he established and ran a laboratory, taught popular courses (including one on "Critical Thinking"), edited *Icarus* (which he turned into the leading scientific journal of solar system studies), supervised graduate students, and maintained a prodigious output of publications. He authored or co-authored two dozen books and more than a hundred scientific papers, many which were seminal, including forty on planetary atmospheres, fifty on other solar system topics, thirty-three on astrophysical and laboratory syntheses of organic molecules, thirty on extraterrestrial biology and SETI (the search for extraterrestrial intelligence), and others on science policy.

At Cornell, Sagan directed an extensive series of laboratory experiments to simulate the atmospheric and surface chemistry of planets, moons, and comets. The results showed that, under a wide range of observed conditions in the solar system, prevailing sources of energy (such as ultraviolet light and electrical discharge) will stimulate the production of complex organic molecules, including the chemical building blocks of life, in high yields. These results were regarded with some skepticism at the time. Today, we know that such substances exist in giant interstellar clouds and on the surfaces of many worlds in the outer solar system. The stuff of life appears to be common in the universe. Sagan assumed that life itself was also widespread.

But what about intelligent life? And advanced civilizations? Despite confident assertions on all sides of the question, no one knows whether they are numerous, rare, or nonexistent. One point however seems clear: Other things being equal, we should expect that the number of advanced civilizations in the universe will be proportional to their average lifetime. If the average civilization lasts no more than a few centuries, then at any given time there will not be very many of them. But if some survive for many millions of years, they will be more

common. In that case, the nearest civilizations might be close enough for us to detect with radio telescopes. The only way to find out is to make the necessary observations. With that in mind, Sagan took part in and worked to build public and institutional support for a number of SETI projects.

As the nuclear arms race began to escalate again in the late 1970s, Sagan became increasingly concerned about the life expectancy of our own civilization. In March 1983, he very nearly died during a ten-hour emergency operation to replace his esophagus. While still in intensive care, he learned about President Reagan's call to build a space-based anti-missile "shield." This he regarded as a technically hopeless scheme that would destabilize nuclear deterrence and perhaps lead to the very war it was supposed to prevent. From his hospital bed, Sagan promptly drafted a petition to Congress opposing the project. Many leading American scientists signed the petition, and Sagan remained a strong critic of "missile defense."

In the same year Sagan also participated in an extensive scientific study of the atmospheric consequences of nuclear war. He and his colleagues calculated that smoke from firestorms in cities might reach the stratosphere and block enough sunlight to cool the Earth, causing a catastrophic "nuclear winter." Their analysis used techniques previously developed to model the cooling of the Earth resulting from major volcanic eruptions and the more drastic cooling due to dust lofted by the asteroid impact that destroyed the dinosaurs. Nuclear winter was at once plausible and controversial. Later, more detailed studies suggested that the climatic consequences of nuclear war would be less severe than calculated, but still sufficient to cripple agriculture in the northern hemisphere.

The widespread discussion of nuclear winter contributed to a substantial rethinking of nuclear war doctrines, particularly in the Soviet Union. In 1986, Sagan briefed the Soviet Central Committee on the subject. Some of those present later said that his effect was profound. Gorbachev personally told Sagan that he had studied the nuclear winter research and it bolstered the case for deep cuts in the nuclear arsenals. Some Russian scientific colleagues credit Sagan with having a major influence on ending the cold war.

Sagan regarded the prevalence of scientific ignorance in a technological society as a prescription for disaster. To promote public understanding of and support for science, he created the popular Cosmos television series, co-founded The Planetary Society (a non-profit public interest group) and used countless articles and interviews to popularize the values of reason, curiosity, critical thinking, and an unbiased search for the truth. While sharply critical of pseudoscience, nationalism, chauvinism, fundamentalism, and other irrational beliefs, he consistently defended the widest freedom of thought and expression. Sagan was never afraid to entertain extraordinary ideas, but he always insisted that "extraordinary claims require extraordinary evidence."

He taught that the unrivalled success of science is due to its combination of openness to new ideas with the obligation to subject those ideas to the most critical scrutiny. He believed that science and democracy share essential values: a free exchange of ideas and information, accountability, and the questioning of authority. Sagan resolutely accepted the verdicts of science even when they contradicted his own fondest hopes and expectations. While he would have liked nothing better than to find evidence for life on other worlds, he instead argued the case that the surface of Venus must be lifeless and that seasonal changes on Mars have nothing to do with life. In both instances, he was guided by a hard-headed analysis of the evidence. By his life and works, he taught that we must always follow the evidence rather than accept uncritically what we merely wish to believe. He maintained that this principle is as valid in the social and political worlds as in the sciences.

In December 1996, after a courageous two year struggle for life, Sagan died of a rare bone marrow disease. The Federation of American Scientists issued a tribute, which noted that in the midst of a life dedicated to scholarship and the popularization of science, Carl Sagan "found the time, and had the courage, to be an intellectual gladiator on issues involving the planet's survival and, in particular, on the prevention of nuclear war." Like a magnificent comet, he illuminated the lives of millions, and we will not see his like again.

—Steven Soter

Section Six: Technical Frontiers

The Goldstone 85-foot antenna, Goldstone, California.

Introduction Steven Soter

From the development of the telescope four centuries ago, technical innovations have brought advances in our understanding of the cosmos. Robotic space probes, orbiting observatories, sensitive photodetectors, adaptive optics, computer modeling, and other technical innovations continue to open new windows on the cosmos.

Nearly all of what we know about the universe comes to us in the form of light, and most of the light reaching the Earth's surface is visible light and radio waves. The other kinds of light— gamma rays, X-rays, ultraviolet, infrared, and microwaves—are largely absorbed by the atmosphere. The only way to observe the universe in those parts of the electromagnetic spectrum is to put telescopes in space. We have now done that, and the opening of each new spectral window on the universe has revealed previously unknown phenomena, from the mysterious distant gamma ray bursts to disks of dust in orbit around young stars.

Telescopes are "light-buckets" designed to gather and concentrate the faint light from distant astronomical sources to reveal fine detail. The larger the telescope, the more light it collects and the more detail it resolves. But visible light images from a telescope on the ground suffer from distortions due to turbulence in the air. One way to solve this problem is to put the telescope in space, as was done with the Hubble Space Telescope. A newer solution is to design a ground-based telescope with adaptive optics, in which the shape of the telescope mirror is adjusted dozens of times per second by a computer to compensate for the changing atmospheric distortions. Adaptive optics will soon allow large telescopes on the ground to produce images as sharp as those from the Hubble Space Telescope.

When the light is clearly focused by a telescope, the observer must record it. In the early days, astronomers drew what they saw. After the invention of photography, they used photographic film. But even the most sensitive film responds to only about two percent of the photons falling on it. Astronomers now prefer the charged-coupled device (CCD), an inexpensive silicon wafer divided into tiny light-sensitive squares called pixels. When the image from the telescope falls on the CCD, each pixel accumulates electrons in proportion to the number of photons falling on it. CCDs typically count more than half the photons and are thus far more sensitive than photographic film.

Astronomers are now exploring the universe using the entire electromagnetic spectrum. They are also going beyond the various kinds of light, by observing cosmic rays and neutrinos from space. Underground observatories detect the elusive and perhaps massless neutrinos produced by nuclear reactions deep in the core of the Sun and other stars. Bursts of neutrinos are also produced by supernova explosions, and one such burst was recently detected. Beyond light and particles, physicists are now developing ground- and space-based observatories for the detection of gravity waves, which are ripples in the very fabric of space expected from the violent merging together of neutron stars and black holes.

Nearer to home, the robotic exploration of the solar system continues apace, with a succession of spacecraft missions—flybys, atmospheric entry probes, orbiters, and landers. Sample return missions to the planets, satellites, asteroids, and comets are being planned. And new laboratory techniques are being used to analyze those extraterrestrial samples delivered to us naturally from space.

Micrometeorites and meteorites from comets, asteroids, and Mars are yielding clues to the early history of the solar system.

During the last four hundred years, Galileo's simple telescope has evolved and diversified into a range of powerful and sensitive instruments on Earth and in space. These technical advances have yielded a rich harvest of discoveries. What new tools will be developed in the next four hundred years, and what secrets of the universe will they reveal? 🪐

To explore some of the technical frontiers in astronomy, we pose the following questions:

What are gamma ray bursts and how are they observed?

Bohdan Paczynski, Lyman Spitzer Jr. Professor of Astrophysics at Princeton University, describes how rapid communication between space-based gamma ray detectors and ground-based optical telescopes are tracking down these mysterious short-lived events.

How can astronomers simulate the processes they cannot study in the laboratory?

Mordecai-Mark Mac Low, Assistant Curator of Astrophysics at the American Museum of Natural History and Adjunct Assistant Professor of Astronomy and Astrophysics at Columbia University, is one of the leading practitioners of astrophysical computer modeling. His essay describes a wide range of computer simulations of complex dynamical events in the universe.

What is NASA's Origins Program?

Alan Dressler, an Astronomer at the Observatories of the Carnegie Institution, in Pasadena, California, describes some of the cutting-edge technologies being developed to answer the ancient questions "How did we get here?" and "Are we alone?"

What new tools are being developed in SETI, the Search for Extraterrestrial Intelligence?

Jill C. Tarter, Director for SETI Research at the SETI Institute in Mountain View, California, describes the development of advanced radio antenna arrays and optical systems to search for possible intelligent signals from space.

Gamma Ray Bursts

Bohdan Paczynski

Cosmic gamma ray bursts are among the most spectacular and mysterious phenomena in the universe. A new burst is detected once a day, on average, and each comes from another direction in the sky. Most of their energy is in very "hard" gamma ray photons ("particles" of light), meaning that each carries about a million times more energy than a photon of visible light. Gamma rays are dangerous, but fortunately for us, the Earth's atmosphere absorbs them. Because of this atmospheric shielding, we can detect cosmic gamma rays only by using instruments in space.

Bohdan Paczynski is Lyman Spitzer Jr. Professor of Astrophysics at Princeton University.

The discovery of the bursts has an interesting history. In 1963, a treaty banning nuclear tests in space was signed. In order to monitor Soviet compliance with the treaty, the U.S. military launched a series of satellites named Vela with detectors sensitive to gamma rays. A nuclear explosion creates an extremely hot fireball which, for a brief time, becomes an intense source of gamma rays. The detectors on board the military satellites were designed to look for such bursts. And the bursts were discovered! However, they were coming not from the Earth but from different directions in the sky, as reported by R. W. Klebesadel and his associates in 1973.

The discovery became headline news for astronomers and physicists, who advanced many speculative theories to explain them. But it was clear that no theory made much sense. Within several years, the general scientific community got tired of this mystery. A few dozen astrophysicists who remained interested reached a consensus: The bursts must be related to some violent phenomena on old, nearby neutron stars, and their typical distance was believed to be several hundred light-years from us, within the disk of our Milky Way Galaxy. The exact cause of the bursts was not agreed upon, with "starquakes," natural nuclear explosions, cometary impacts, and many other hypothetical processes proposed at different times. A few dissidents, including the author, argued that the bursters may in fact be billions of light-years away, at cosmological distances (far outside our Galaxy), but this view was not taken seriously. While a small number of scientists argued about the physics of the bursts, the rest did not care much, and no popular textbooks even mentioned gamma ray bursts.

The views changed dramatically when G. J. Fishman, C. A. Meegan and their associates presented the results from BATSE (Burst And Transient Source Experiment), one of the

instruments on the Compton Gamma Ray Observatory, launched by NASA in 1991. The BATSE large area detectors were much more sensitive than anything previously available, and thus were able to detect many more weak bursts. Also, with eight detectors located at the eight corners of the satellite, BATSE could determine the location of each burst with an accuracy of several degrees, adequate to provide useful information about the overall distribution of sources over the sky.

Figure 1 shows several examples of BATSE bursts. A huge diversity of time variability is readily apparent. The shortest burst ever recorded lasted less than 0.01 second, the longest remained active for over ten minutes. Some showed smooth variations over time, others varied rapidly with a rich time structure. Note that the intensity of gamma ray emission as presented in each panel is the total emission from all gamma ray sources in the whole sky, since everything is recorded by the BATSE detectors. Yet when a burst is active, it shines much brighter than all the steady gamma ray sources in the sky combined. The contribution of the rest of the sky, called the background, is seen before and after every burst. The background is approximately eight to ten units in Figure 1, while the strongest of the bursts shown, number 249, has a peak intensity of over 200 units, or about twenty times brighter than the rest of the sky. Some bursts are even much stronger than this one. The weakest, like number 408 in Figure 1, are only about as strong as the background.

Figure 2 shows the distribution of the first 2,000 recorded BATSE bursts over the sky in galactic coordinates. The direction to the center of our Galaxy is at the center of the map, and the galactic equator is the horizontal line through the middle of the map. The spatial distribution of gamma ray sources looks random in all directions. Many statistical tests

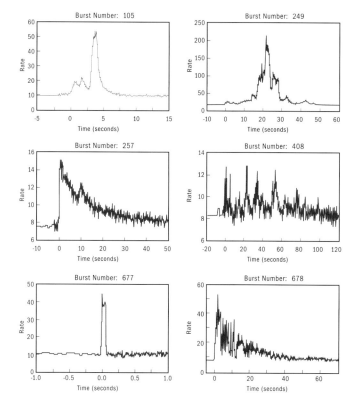

Figure 1: Six examples of time variability of gamma ray bursts detected by BATSE. Among these bursts the shortest (number 677) lasted only 0.1 second, the longest (number 408) lasted for two minutes. Note that all the constant gamma ray sources visible in the sky contribute to the "background" intensity seen preceding and following each burst. The rate (vertical axis) is marked in thousands of gamma rays, counted per second.

provide this information within several hours of the burst event. This allowed the ground-based optical and radio observers to look for possible transient optical and radio afterglows that might follow the gamma ray bursts. In May 1997, H. Bond used a small telescope at Kitt Peak to discover a variable optical source at the location of Gamma Ray Burst (GRB) 970508. Following this discovery a team of Caltech astronomers obtained its spectrum with the huge Keck Telescope in Hawaii. For the optical transient they found the measured redshift to be $z = 0.835$, which means that the observed light waves are all 83.5 percent longer than those emitted at the source. This cosmic redshift is due to the expansion of space, and translates to a distance of approximately 7 billion light-years away from us.

During 1998 and 1999, BeppoSAX provided good coordinates of bursts approximately once a month, and more than a dozen optical and/or radio afterglows have been detected. Redshifts were measured for ten of them, ranging from $z = 0.45$ to $z = 3.4$, roughly the distance range typical for quasars. In most cases, faint host galaxies were discovered within a fraction of a second of arc from the afterglow location, with some evidence that the bursts are located in star-forming regions.

What does it all mean? Interestingly, the afterglows were theoretically predicted prior to their observational discovery. The reasoning was very simple. Every powerful explosion we observe in the universe is followed by an extended emission from a hot, low-density gas located near the explosion. We observe this

demonstrated that this visual impression is correct, and there is no concentration either towards the galactic equator or the galactic center, as we would expect if the sources were associated with our Galaxy. Preliminary BATSE data, presented at a scientific conference in September 1991, instantly convinced about half of those present that gamma ray bursts cannot be associated with our Galaxy, and must therefore come to us from cosmological distances. But the other half were not persuaded, mainly because bursts observed from so far away would require enormous amounts of energy to appear so bright.

The breakthrough was achieved thanks to a new Dutch-Italian satellite named BeppoSAX. It was capable of measuring the location of the gamma ray bursts with an accuracy of about three minutes of arc (about ten percent the angular size of the Moon on the sky), and to

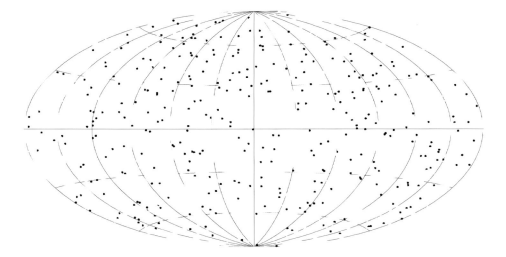

Figure 2: The distribution of the first 2,000 gamma ray bursts detected by BATSE is shown in galactic coordinates. The center of our Milky Way Galaxy is in the middle of the figure, while the plane of the Galaxy is the straight horizontal line cutting the figure in half. The gamma ray sources are clearly distributed randomly over the sky, suggesting that they are at cosmological distances rather than within the Milky Way. This means that the gamma ray sources must be extremely powerful to appear so bright at such distances.

around supernovas, and also around quasars, so it was natural to expect something similar around the bursters. Why is it so natural? There is no perfect vacuum anywhere in the universe. There is always some gas between the stars, and even between galaxies. When the debris ejected by an explosion plows into that tenuous gas, both are heated, and as a consequence they radiate. The details are complicated, but the general principle is simple. The cause of the explosion doesn't matter, only how much energy it has. In the case of gamma ray bursts, the amount of energy released in just a few seconds is about a hundred times greater than the total energy our Sun will radiate during its entire lifetime of 10 billion years. Some bursts are even more powerful than that. Therefore, some kind of an afterglow was expected to follow such a powerful explosion.

What is the nature of gamma ray burst explosions? What are the objects that explode? What is the energy source? The simple answer to these questions is: we don't know. The location of the afterglows near or within star-forming regions suggests that gamma ray bursts may be related to so-called Type II supernovas. These are explosions of stars at least ten times more massive than our Sun. Every massive star runs out of nuclear fuel within several million years of its birth. That isn't long enough for a star to move very far, so it dies close to its birthplace, i.e. near a star-forming region.

When the core of a massive star runs out of nuclear fuel, it collapses and forms a hot neutron star, only about twelve miles in diameter, but with a mass fifty percent greater than our Sun. It is very hot indeed, with its heat generated by the catastrophic gravitational collapse. Within ten to twenty seconds, 99.7 percent of this energy is radiated away as a powerful burst of invisible neutrinos. Only 0.3 percent is transferred to the outer layers of the star, where it generates a spectacular stellar explosion that can be seen across the universe. The strong optical emission from such a supernova explosion lasts for about a month or so, and it may outshine a whole galaxy during that time. Yet, 99.7 percent of the huge energy was carried away from the star with the elusive neutrinos.

The optical light accompanying a gamma ray burst carries about a hundred times more energy than the light from a supernova. But, as we saw, most of the energy of a supernova is "wasted" in invisible neutrinos. It cannot contribute to the observed optical and gamma ray light associated with gamma ray bursts. So, how can we make an explosion efficient enough to account for gamma ray bursts?

A popular concept is to tap the enormous energy stored in the form of stellar rotation. A single massive star sheds most of its rotation in an immense stellar wind when it becomes a red supergiant, prior to its core collapse. However, if a massive star orbits a close companion star with a period of several hours, then tidal gravitational forces (like those by which the Moon raises tides on the Earth) keep the massive star rapidly rotating. When its core then collapses, the conservation of angular momentum causes the resulting neutron star to rotate extremely fast (up to a thousand times per second). All the heat energy is rapidly taken away by a powerful neutrino burst, just as in a normal supernova, but the enormous rotational energy remains. If the star has very strong magnetic fields, then it may be possible to transfer most of the rotational energy into an explosion powerful enough to make a gamma ray burst. But, the word "if" appears twice in this paragraph, which means that we cannot be sure.

I am confident that the solution to the puzzle will be found with new unexpected observations, and new theoretical insights. A spectacular example of a new type of observation was provided on January 23, 1999 by the ROTSE (Robotic Optical Transient Search Experiment) team led by C. Akerlof. The instrument was very small, made of four telephoto lenses, each with its own CCD camera, a sensitive electronic detector similar to those in camcorders and digital cameras. During the gamma ray burst GRB990123, the computer analyzing BATSE data in real time sent the coordinates of the ongoing burst to the computer controlling the ROTSE camera system. The latter took multiple images of a large region in the sky around the approximate location of the burst. A few hours later the analysis of ROTSE images revealed a bright optical flash, lasting several minutes and brighter than ninth magnitude at its peak. A spectrum of this fading optical afterglow taken with several big telescopes revealed that it had a cosmic redshift corresponding to a distance of over 10 billion light-years. Yet, the optical flash was so bright that you could have seen it through ordinary binoculars if you knew when and where to look!

Let us imagine for a moment that the burst GRB990123 was not at the other end of the universe, but within our Galaxy at a "small" distance of three thousand light-years away (for comparison, the nearest stars are only a few light-years away). Its optical flash would then be as bright as the Sun at high noon! And the gamma ray pulse would be several thousand times more intense than that. This may sound dangerous, but the bursts are so rare that we may have to wait billions of years for any of them to occur so close to us.

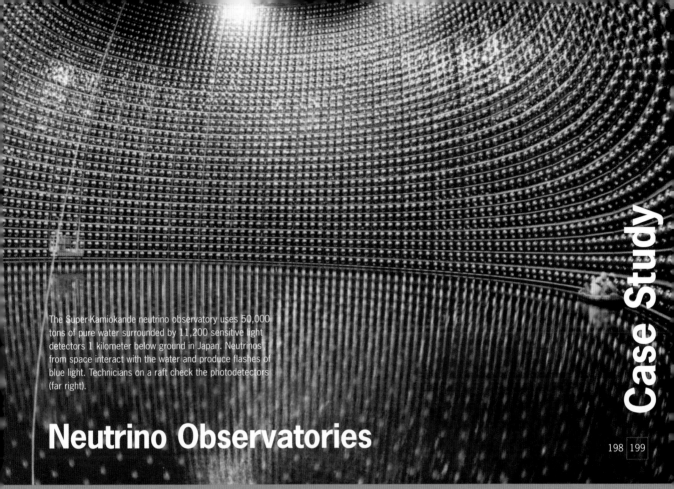

The Super-Kamiokande neutrino observatory uses 50,000 tons of pure water surrounded by 11,200 sensitive light detectors 1 kilometer below ground in Japan. Neutrinos from space interact with the water and produce flashes of blue light. Technicians on a raft check the photodetectors (far right).

Neutrino Observatories

The image of an astronomer peering through the eyepiece of an enormous telescope in a mountaintop dome late at night is still a stereotype in popular culture. While this sort of thing was, in fact, common a generation ago, astronomers today rarely "look" through telescopes. Instead they look at computer monitors while directing the operation of sensitive electronic light detectors attached to telescopes. The astronomer does not even have to be anywhere near the telescope, which could, for example, be in space.

A wide range of modern detectors can "see" not only visible light from celestial objects but also every other kind of light in the electromagnetic spectrum, from radio waves to gamma rays, all invisible to human eyes. However, light is not the only carrier of information across space.

Subatomic particles called neutrinos permeate the universe. In fact, during the next second, trillions of neutrinos from cosmic sources will pass through your body with no effect. Like visible light from stars, radio waves from galaxies, and X-rays from matter spiraling down black holes, neutrinos can also reveal something of the cosmos.

Neutrinos entered theoretical physics in 1930 as a way to understand beta decay, the process by which a radioactive atomic nucleus spits out an electron. Experiments showed that the total energy of the nucleus plus the ejected electron after the decay was less than the energy of the initial nucleus. This appeared to violate the conservation of energy, a fundamental principle of physics which says that energy is neither created nor destroyed.

In 1987, the most spectacular supernova seen in four centuries appeared in the Large Magellanic Cloud, a satellite galaxy of the Milky Way. The exploding star released a burst of neutrinos observed on Earth. This composite image made in 1994–97 shows rings of gas expanding from the dying star. Hubble Space Telescope.

The Austrian physicist Wolfgang Pauli, so convinced of the conservation principle, made a daring intellectual leap by proposing that an unknown particle carries off the missing energy. To agree with the observations, this particle had to be electrically neutral, possess practically zero mass, and move at the speed of light. Pauli suggested this in a letter to his colleagues but didn't publish it. He understood that a scientific theory is almost worthless unless it can be tested by observation or experiment, and he was concerned that such particles could

never be detected. Then in 1932, James Chadwick discovered the neutron, a particle with nearly the same mass as the proton but no electric charge. Pauli gained confidence and published his idea. The physicist Enrico Fermi named Pauli's particle the neutrino, meaning "little neutral one" in Italian.

The hypothetical neutrino served to explain beta decay, and even led Fermi to recognize that there must be a new nuclear force, now called the weak force, if neutrinos actually exist. But for years these ideas remained unconfirmed. Other physicists would poke fun at Pauli, some even calling the neutrino "the little neutral one who isn't there."

The task remained to detect these things, and that would be very difficult. Neutrinos have almost no interaction with matter. In theory, a high-energy neutrino could pass through a hundred light-years of solid steel without hindrance. And radioactive atoms, the only source then known for the hypothetical neutrinos, should emit relatively few of them. The picture changed in the 1950s, however, with the advent of nuclear reactors. Such power plants would produce a flood of neutrinos if Pauli and Fermi were correct.

Clyde Cowan and Frederick Reines decided to use a nuclear reactor to search for neutrinos. They designed a 10-ton detector to record the tiny spark of light expected from the rare interaction of a passing neutrino with a proton inside the device. They placed their detector next to the newly-completed Savannah River nuclear power plant in South Carolina. Day after day, they recorded data and slowly gathered evidence from the occasional flashes of light. After a few years, Cowan and Reines had enough data to prove that neutrinos actually exist. Pauli was naturally ecstatic and made sure his colleagues knew about the result.

By this time, physicists realized that the nuclear reactions that power the stars must also produce neutrinos in enormous numbers. Raymond Davis decided to try observing these cosmic neutrinos. But in order to do so, he would have to shield his detector from cosmic rays, the high-energy charged particles constantly bombarding the Earth from space. His solution was to place the detector deep underground. Cosmic rays cannot penetrate hundreds of meters of rock, but for neutrinos the entire Earth would be transparent. In 1967, Davis installed a large tank of cleaning fluid in a deep gold mine in South Dakota. Any neutrino passing through this tank had a tiny probability of hitting a chlorine atom in the cleaning fluid and converting it to a radioactive argon atom, which could easily be detected. This experiment found about one neutrino every few days. They were attributed to the Sun.

Neutrino observatories around the world have continued to collect solar neutrinos. According to astrophysical theory, the Sun should emit about two percent of its energy in the form of neutrinos. But the number of neutrinos actually detected on Earth is only about a third as many as predicted by the theory. Perhaps there is something we don't understand about the Sun or about neutrinos. The problem of the "missing solar neutrinos" continues to occupy the attention of many investigators.

In 1987, neutrino astronomy made a stunning and unexpected advance. The two most sensitive neutrino observatories in the world, one in Japan and the other in Ohio, had recently become operational when the first naked eye supernova since 1604 blazed forth in the sky. That event, called SN1987A, announced the explosive death of a massive star, which radiated as much visible light during one day as the entire Milky Way Galaxy. Such an explosion ought to produce a strong burst of neutrinos.

The supernova was in the Large Magellanic Cloud, a companion galaxy to our Milky Way. It was just close enough for the enormous dose of neutrinos passing through the Earth to trigger a response in the two neutrino observatories.

Both observatories detected a 12-second burst of neutrinos about three hours before the supernova became optically visible. The explosion in the core of the star produced the neutrinos which raced through the overlying stellar mass at the speed of light and out into space. Three hours later the shock wave from the core reached the surface of the star and blew it apart to make the visible supernova explosion. No one could doubt that the observed burst of neutrinos came from SN1987A. The number of neutrinos detected by these two observatories allowed physicists to estimate how much matter was involved in nuclear reactions during the supernova explosion.

This event was a watershed in neutrino experiments and has prompted the construction of more sensitive detectors, all placed deep below the Earth's surface. These mineshaft observatories, a far cry from mountaintop optical telescopes, should reveal more about the nature of supernovas and other high energy cosmic events, as well as provide more clues about the elusive neutrinos.

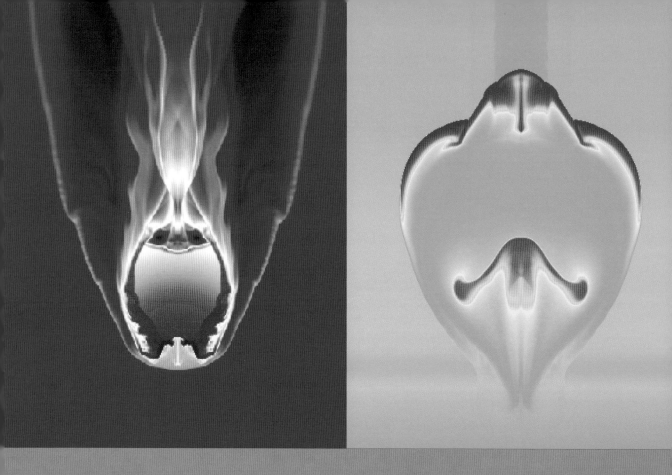

Astrophysical Computer Modeling

Mordecai-Mark Mac Low

A half-mile (800 meters) wide pile of gravel and dirty snow
hurtles by Jupiter, barely evading the grasp of its gravity. The
feather-light difference in Jupiter's gravitational force between
the pile's near and far sides strews it out over distances greater
than the separation of the Earth from the Moon. Over the next
year, the bits of rubble from the ill-fated comet (for that is what
it was) gravitationally reassemble themselves into a chain of
twenty smaller objects. Jupiter's gravity eventually asserts itself,
drawing the whole chain back towards the planet on an even

Mordecai-Mark Mac Low is Assistant Curator of Astrophysics at the American Museum of Natural History,
as well as Adjunct Assistant Professor of Astronomy and Astrophysics at Columbia University.

more perilous orbit as Earth-bound astronomers finally notice its desperate circumstances.

With the whole world watching, the first of the fragments plunges into Jupiter's atmosphere at over 100,000 miles per hour (45 kilometers per second), becoming the mother of all meteor strikes. Penetrating below the ammonia clouds of the giant planet, the white-hot heat from its own shock wave vaporizes the comet fragment, releasing the energy equivalent of thousands of nuclear bombs. The resulting fireball would stretch from New York to Chicago. It blows material clear out of Jupiter's atmosphere, but is nevertheless a mere pinprick to the planet.

To understand the violent fate of Comet Shoemaker-Levy 9 in July 1994, astronomers turned not just to large telescopes and traditional mathematical calculations, but also to computer models of the encounter. In addition to observation and mathematical theory, computer modeling has emerged in the last half-century as a third way of understanding our universe. Driven by the astonishing increase in speed and capability

of computers, computational modeling is based on mathematical descriptions of physical phenomena, but then borrows many of the techniques of observation to understand its results.

In the case of the comet, the breakup into fragments was modeled using one of the most fundamental mathematical descriptions of the physical universe: the law of gravity. The orbits of thousands of particles placed initially in the same half-mile wide sphere were computed as they raced through the gravitational fields of Jupiter and the Sun, as shown in Figure 1. Mathematical computations of orbits under the influence of a planet and the Sun draw on a

Figure 1: Three stages in the breakup of Comet Shoemaker-Levy 9 into a string of fragments. Tidal forces from Jupiter are tearing apart the comet. The particles used in the computation are shown as circles or points, and their directions of motion are shown in the first panel as arrows. The first panel shows the shape of the parent body as it passes closest to Jupiter, while the later panels show the distribution of fragments after nearly two and ten hours. The scale of the first panel is only one-tenth the scale of the second two panels. Computation by E. Asphaug & W. Benz.

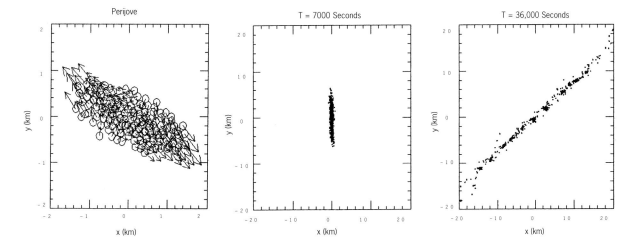

body of work stretching back over 300 years to Sir Isaac Newton's formulation of the law of gravity. In the nineteenth and early twentieth centuries, orbit calculations were already carried out by "computers," but the word then referred to a profession rather than a machine—an arduous profession that offered one of the only ways then available for women to participate in astronomical research. Although single orbits could then be computed by hand, the computation of thousands of orbits for a model of a fragmenting and reassembling comet was totally impractical. Now, it can be done in an hour or two by a single astronomer with a computer, allowing her to make many different models and choose the best by comparison with the observations.

If we let the particles represent not boulders but stars (or really groups of tens of thousands of stars), simultaneous computations of tens of thousands of particle orbits can allow us to understand why galaxies have their beautiful spiral arms. Placing the particles in an initially uniform, rotating disk and computing their interactions shows that spiral arms form whenever the disk is even slightly perturbed, as a natural result of the gravitational interactions between the stars, as shown in Figure 2.

Computations of the collisions of two disk galaxies show that the same mutual gravitational interactions disrupt the orderly disks and throw stars out in spectacular streamers tens of thousands of light-years long, as shown in Figure 3.

These same techniques have led to a revolution in our understanding of the large-scale structure of our universe and the formation of galaxies. In the earliest centuries after the Big Bang, matter in the newborn universe was uniformly distributed. Over time, regions of slightly greater density attracted more and more mass, eventually forming stars and galaxies, while regions of slightly smaller density were emptied out, forming cosmic voids that remain today.

Figure 2: Model of two interacting spiral galaxies showing how spiral arms are naturally formed by small perturbations to galactic disks. Each particle used in the computation is indicated by a point in the image. Computation by J. Barnes.

Figure 3: Interacting galaxies at a later time after they have been severely distorted by their mutual gravitational attraction. Similarly distorted galaxies are actually observed. Computation by J. Barnes.

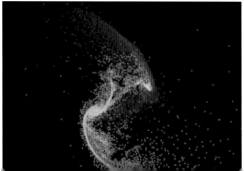

HD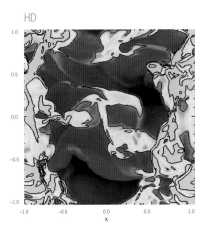

MHD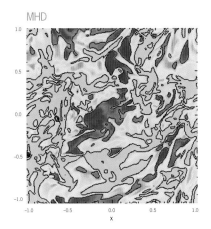

Figure 4: Slices through models of supersonic turbulence in star-forming clouds without (HD) and with (MHD) magnetic feilds. Stars can form in the densest red regions. These images are roughly one light-year across. Computation by F. Heitsch and M.M. MacLow.

Models of this process begin with not tens of thousands but tens of millions of particles, almost uniformly distributed through a representative region of the universe. Each particle now represents the mass of millions of stars, still a small fraction of a galaxy. The orbits of these particles under the mutual influence of all the other particles in the region lead to the formation of the cosmic web, as shown in Figure 7.

Returning to the case of the comet on its fateful final orbit, we find that computations including only the force of gravity accurately predict its behavior only as long as no other forces are important. However, on this orbit, the comet's path crosses within one Jupiter radius from the center of the planet, ensuring a collision. When the comet fragment screams into the wispy outer reaches of the planet's atmosphere, pressure begins to act on it, and we enter the realm of hypersonic gas dynamics. Although the basic laws of gas flow had already been described 300 years ago by Euler, it was not until the middle of the twentieth century that the

properties of strong shock waves and massive explosions were carefully computed, for application to the development of rockets and nuclear bombs. Even to this day, the U.S. nuclear weapons development program maintains the largest computers available for such computations, consisting of thousands of individual processors, although open scientific research centers are in strong competition.

Gas pressure varies over space, so to compute its effects, we define a grid throughout the region of interest, like laying down a sheet of graph paper. At each point of the grid, we store the current local gas pressure, density, and velocity. In the case of the comet collision, the grid covered a small region in Jupiter's atmosphere, with a sphere (representing the comet) traveling at many times the speed of sound near the top. We then compute how the gas at each point will interact with its neighbors over a very short step in time—gas will flow away from regions of high pressure and toward regions of low pressure. Repeating this for thousands of time steps, we can follow the gas flow. The entering comet fragment drives a high-pressure shockwave in front of it, as shown in Figure 5a, while the rest of the atmosphere remains undisturbed until the shock wave hits it, heats it, and starts the expansion of the explosive blast shown in Figure 5b.

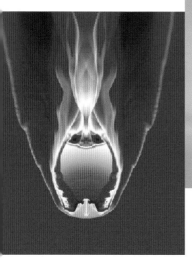

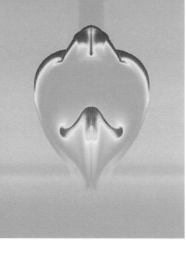

Figure 5a (left): Slice through a model of the entry of a fragment of Comet Shoemaker-Levy 9 falling through the atmosphere of Jupiter. The gas behind the bow shock reaches temperatures over 30,000 degrees. This figure shows a region six miles (10 km) across. Computation by M.-M. Mac Low & K. Zahnle.

Figure 5b (left): Slice through a model of the explosion resulting from the impact of a fragment of Comet Shoemaker-Levy 9. This model shows a region 600 miles (1,000 km) across. The visible cloud layer on Jupiter is at the level of the blue stripe near the bottom of the image. Computation by M.-M. Mac Low & K. Zahnle.

These computations of the comet impact were inspired by computations that I had done earlier of the effects of supernova explosions on the interstellar gas in the disks of galaxies. It turns out that, although the distances are tens of trillions times larger, and the explosion energies even more extreme, the physical mechanisms are quite similar. I basically scaled down numerical models that I had computed of the huge cavities of hot gas called superbubbles, formed by multiple supernova explosions in the galactic disk. I reduced the distances down from light-years to miles, by some thirteen powers of ten (1 followed by 13 zeros), and the explosion energies by even more, a factor of twenty-three powers of ten, in order to find virtually the same expansion of an explosion in a stratified atmosphere, as shown in Figure 6.

Models including the effect of both gas dynamics and gravity have proven to be useful not just in understanding the deaths of stars in supernova explosions, but also the birth of stars in clouds of gas and dust. These clouds are opaque to visible light, creating the rifts and gaps that can be seen in the band of the Milky Way on a dark night. They are millions of times denser than the rest of the interstellar gas, but nevertheless have densities comparable to the best laboratory vacuums here on Earth. Because they extend over light-years, however, their total

mass can be millions of times the mass of the Sun, providing the raw material for many stars. In the absence of any restraining forces, gravity would cause these clouds to collapse into stars in less than a million years—an astrophysical blink of an eye. However, the pressure and turbulent motions of the gas resist the collapse, aided by magnetic fields threading the gas.

Modeling this interplay of forces using grid and particle methods allows us to understand how gas can go from near-vacuum densities and near absolute zero temperatures to the million-degree ionized gas, denser than rock, in the center of a star. One current scenario is that shock waves from supersonic turbulence can prevent large regions of gas with hundreds of solar masses from collapsing, but at the same time sweep up enough gas to promote the formation of small numbers of stars. A model of this process is shown in Figure 4.

How will astrophysical computer modeling develop in the next decade? The capability to do straightforward 3D gas dynamics at moderate resolution has only developed recently. Now the challenge is twofold: first, we need to include more of the relevant physical processes, including the chemical behavior of the gas as it is heated to millions of degrees by hypersonic shock waves and then cools down close to

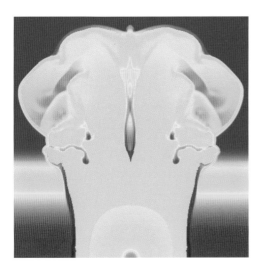

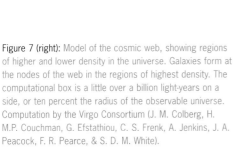

Figure 6 (left): Slice through a model of the explosion of multiple supernovae in the plane of a spiral galaxy. The plane of the galaxy is at the bottom of the image and perpendicular to the plane of the page, and only the part above the plane of the page is shown. This image is 3,000 light-years across. Computation by M.-M. Mac Low, R. McCray, and M. L. Norman.

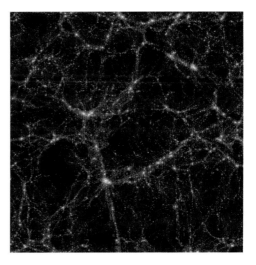

Figure 7 (right): Model of the cosmic web, showing regions of higher and lower density in the universe. Galaxies form at the nodes of the web in the regions of highest density. The computational box is a little over a billion light-years on a side, or ten percent the radius of the observable universe. Computation by the Virgo Consortium (J. M. Colberg, H. M.P. Couchman, G. Efstathiou, C. S. Frenk, A. Jenkins, J. A. Peacock, F. R. Pearce, & S. D. M. White).

absolute zero. Second, we need to extend the resolution of our models to include a greater range of size scales, so we can follow individual galaxies in cosmological simulations, or individual stars in star formation simulations. One method to reach greater resolution is to automatically select regions of interest and insert higher-resolution grids in those regions, subdividing repeatedly when necessary to go to very small scales.

The new generation of supercomputers presents its own challenges. Rather than being very fast single processors, they have hundreds or even thousands of off-the-shelf microprocessors tied together by fast networks. Programming these machines requires the coordination of all of these thousands of units without generating traffic jams on the internal networks as bad as those on the

Internet. Another direction being pursued is to perform the most time-consuming parts of simulations directly in specially-designed chips, achieving supercomputer speeds on the desktop for those specific problems. This technique has been especially fruitful for particle simulations such as those described above.

Finally, the never-ending battle to win insight from computation will depend more and more on scientific visualization and simulation of observations. Increasingly, astrophysicists will attempt to understand complex observations with complex numerical simulations, but always with the ultimate aim of deriving a confirmable, predictive description of the universe. ♄

NASA's Origins Program:
Answering the Ancient Questions

Alan Dressler

When the twentieth century began, astronomers were engaged as surveyors, timekeepers, and naturalists describing the heavens. But over the last hundred years, they have assembled an extraordinary physical understanding of what we see in the cosmos and how it works—for example, how stars are born, change through their lifetimes, and die. Something more is emerging, however, at the start of this new century. We have become aware that there is a story out there, one that resonates with the oldest questions of humankind. Although it is still

Alan Dressler is an Astronomer at the Observatories of the Carnegie Institution, in Pasadena, California, the Department of Astronomy.

tangled in and partly obscured by the myriad data we have collected about the universe, it is a story of how we ourselves came to be, of how the universe evolved from a super-hot gas of simple atoms to produce the delicate environments for complex biochemistry. It is the story of our own origins.

Following the stunning success of the Hubble Space Telescope, NASA launched an aggressive research program, called Origins, that seeks to answer the ancient questions "How did we get here?" and "Are we alone?" This is a story of how the hot gas of the Big Bang cooled to form galaxies and, within them, stars with their orbiting planets, and of how the raw materials of the cosmos would eventually give rise to complex chemistry and life. To understand the origins of galaxies, stars, planets, and life—this is the essence of the NASA Origins program.

It is clear that an understanding of how stars and planets form will tell us a lot about how we came to be. But why is the formation of galaxies, in this case, of our own Milky Way Galaxy, an important step in the origin of life? Our planet is a rocky one, made of heavy chemical elements such as iron, silicon, oxygen, and carbon. Ninety percent of our own body weight is also made of heavy atoms. But the box the universe came in was labeled "heavy elements not included!" The Big Bang produced mainly hydrogen and helium—which are not the building blocks of Earth-like planets. Where did all the heavy stuff come from? From stars.

In their cores, stars convert vast quantities of hydrogen and helium into the heavier chemical elements through the process of nuclear fusion.

Starlight is just another product of this fusion. At the end of their lifetimes, stars return some or all of their processed material to space, for example, in supernova explosions, thereby enriching the interstellar gas with heavy chemical elements. And here is where galaxies play a crucial role: if it were not for the powerful gravity of billions of stars gathered in a galaxy, the enriched gas would escape to intergalactic space. Instead, the galaxy retains the gas, and thus grows ever richer in the essential building blocks of planets and life.

When the Milky Way Galaxy was about 5 billion years old, it gave birth to our solar system, from gas that contained heavy elements left over from the life cycles of many generations of stars. Every atom in your body except hydrogen was once at the center of a star. That's what the astronomer Carl Sagan meant when he said "We are star stuff."

But, how exactly did this process of cosmic evolution happen? The first generations of stars were born about half a billion years after the Big Bang, probably in giant star clusters that later coalesced into big galaxies. The production of heavy elements in stars increased dramatically over the next few billion years. In order to fill in the details, astronomers are using telescopes like time machines. By observing starlight that has been traveling to us from distant galaxies for billions of years, we can literally see what the universe was like billions of years ago. In fact, using the Hubble Space Telescope and the largest Earth-based telescopes, we have already looked back to within a few billion years of the Big Bang. We have taken pictures of young galaxies and obtained crude spectra of their light. These tell us how far away the galaxies are and how far back in time we are looking.

But we have not yet been able to make the more difficult measurements that can help answer the crucial questions: how much of the heavy elements had each galaxy collected during its infancy? At what rate were stars born in those early phases? How were the pieces of galaxies assembled? And how massive were the galaxies that could retain the heavy elements needed to make planets and life? We can answer these questions, or at least make a good beginning at it, only by observing more detailed spectra of distant galaxies. That will require building a much more powerful space telescope, now called Next Generation Space Telescope (NGST). It will collect about ten times more light than the Hubble Space Telescope. Present plans call for a mirror 8 meters across (compared to the Hubble's 2.4 meters). But the real advantage of NGST will be its ability to make detailed measurements in the infrared part of the spectrum, where it will be thousands of times more sensitive than any telescope ever built.

The infrared capability is essential because the light waves emitted by these infant galaxies have been stretched out while traveling for so long across an expanding universe. This journey has shifted the light from the visible into the longer wavelength infrared part of the spectrum. The problem is that, until now, all our telescopes have been "warm," at what we call "room temperature"—hundreds of degrees above absolute zero. This means the telescopes themselves glow brightly in the infrared, so that looking for faint, infrared galaxies with them is like looking for a lit match inside a hot kiln!

The NGST will solve this problem by keeping the telescope very cold, which is possible in space but not here on Earth. Even the Hubble is heated by the warmth of the Earth, which it circles in a low orbit. But the NGST will be in orbit at about twice the distance of the Moon, where it will cool to about fifty degrees above

absolute zero. That is cold enough to see the faint infrared evidence for the actual birth of galaxies in the distant early universe. The first telescope to try out this cryogenic technology will be the Space Infrared Telescope Facility (SIRTF). An early Origins mission, SIRTF will be our most sensitive probe so far of the earliest moments of star birth in the universe. But the detailed study of young galaxies will require the much larger NGST, now scheduled for launch in 2008. Already there is talk of a next-next generation space telescope, even more sensitive, to see how galaxies formed even further back in time, and complementary ground-based optical telescopes with mirrors up to one hundred meters across!

Because SIRTF and NGST will be extraordinarily sensitive to infrared light, they will also make huge steps in another part of the Origins mission—the study of how stars form. Here there are two substantial advantages for a sensitive infrared telescope. First, stars form in cool dusty disks of gas, which emit mostly infrared light. Second, while the abundant dust in these stellar nurseries absorbs almost all the visible light from those stars, it is relatively transparent to the longer wavelength infrared light. Thus we can use infrared light to observe both the dusty disks and the stars forming within them. These dusty gas disks are really the construction material for building stars, and when the process is largely over, they become the dumping ground for the leftover material. These new telescopes will teach us more about the process of star formation, and may also provide our first glimpse of planets growing around newborn stars. It is within these disks that tiny grains begin to coalesce like snowballs, first forming larger and larger grains, then pebbles, and finally rocky and icy planets.

Recently, astronomers have made spectacular progress in detecting planets around other stars. These planets were found indirectly by observing the gravitational "wobble" they produce in the stars around which they orbit (see the essay by David C. Black in Section One). These are giant gaseous planets, more akin to Jupiter than the tiny rocky planet we call home. Still, it was reassuring to know that planets do in fact orbit other stars. Surprisingly, these giant planets are often found closer to their stars than the Earth is to the Sun. This is definitely not where we expected to find icy, gaseous planets. Rather, our theories had predicted that the volatile elements should condense only in the cooler outer zones of a planetary system, as was the case for Jupiter, Saturn, Uranus, and Neptune. One proposed explanation is that the observed extrasolar giant planets originally formed far from their stars but have since migrated in closer to them. If so, they must have incorporated or gravitationally expelled any little Earth-like worlds in their way. We may, in fact, be lucky enough to live in one of the rare solar systems where the giant planets mind their place.

On the other hand, the apparent prevalence of extrasolar systems having giant planets close to stars may be biased by "observational selection." The gravitational wobble method cannot yet detect extrasolar planets as small as the Earth. In that case, the examples discovered so far might not be a very representative sampling of other planetary systems. But even if giant planets orbiting close to their stars turn out to be the rule, many of them could still have moons large enough to support life.

Another NASA space telescope, the Space Interferometry Mission (SIM), may be the first to find an Earth-like planet, when it looks for the small pivoting motion of a star tugged on by such a planet. SIM is scheduled for launch in 2006. The big payoff, however, will come from looking for small planets directly, using the

visible light they reflect from their star, or the infrared light they re-emit from their own heat. You might think the reason we haven't seen such planets is that they are too faint, that our telescopes are just not big enough. Not so. The problem is that optically searching for a planet next to a bright star is like trying to see a firefly while looking into a giant searchlight. However, new telescopes are coming that will be able to block out the light from the star and allow us to see the tiny worlds that circle it.

Interferometric telescopes like the SIM combine the light gathered by two or more separate mirrors to make an image equivalent to that from a single much larger mirror. These telescopes can also combine the light waves "out of phase" so that they interfere, much like waves at the beach can cancel when they meet crest to trough, resulting in a flat ocean for a while. SIM will be the first telescope to test this technology in space. It will open the way for a future space telescope, called the Terrestrial Planet Finder (TPF), which will use lasers and computers to position the distance between four or more orbiting telescopes flying in formation to a precision better than a millionth of an inch. The light waves from these separate mirrors can then be combined to interfere and make a targeted star go dark. By reducing the star's intensity by more than a million, the TPF telescope should reveal the dimmer planets that may orbit the star, be they giant worlds like Jupiter or tiny worlds like Earth.

But that isn't all. The TPF will spread out the incredibly faint light from such planets into a spectrum good enough to detect carbon dioxide, water, and oxygen (in the form of ozone) in the atmosphere of each planet. The presence of water vapor may indicate an Earth-sized world with oceans or lakes. Should we be fortunate enough to find oxygen as well, it would suggest the presence of life on such a

210 211

planet. Because oxygen combines readily with many rocks (in the same way that iron rusts), it must be constantly replaced or it will disappear from the air. Here on Earth, most of the atmospheric oxygen is produced by plants, as part of their respiration, and it is likely that atmospheric oxygen on another world will also be the signal that life, at least microscopic life, is abundant.

After this first step, bigger, more powerful space telescopes will follow, reaching to more distant stars and gathering data about the atmospheres of their planets. Eventually this route should lead either to decisive evidence for life, or good evidence that life as we know it is rare. Eventually, using an interferometer array of giant telescopes separated by thousands of miles in space, we might even observe the surface of another Earth-like world. It's a grand dream, but well worth the enormous challenge.

How life begins is the last remaining piece to the Origins program, and arguably the most exciting scientific adventure at the start of the twenty-first century. We once thought the presence of life required that surface conditions be "just so." We might call them the Goldilocks conditions—not

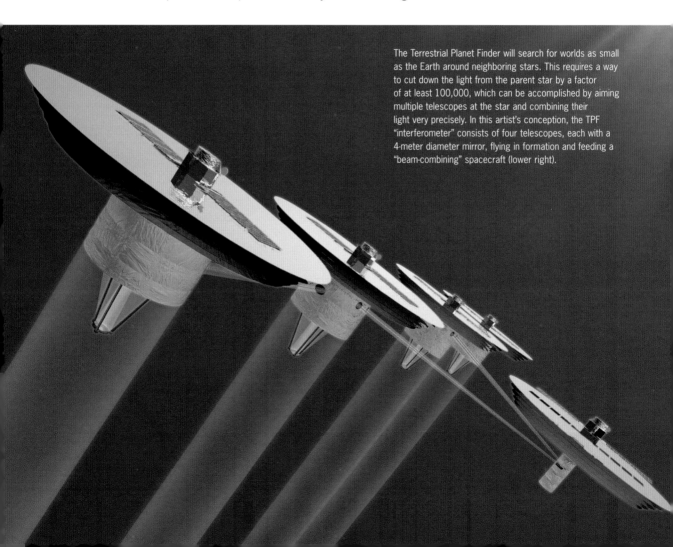

The Terrestrial Planet Finder will search for worlds as small as the Earth around neighboring stars. This requires a way to cut down the light from the parent star by a factor of at least 100,000, which can be accomplished by aiming multiple telescopes at the star and combining their light very precisely. In this artist's conception, the TPF "interferometer" consists of four telescopes, each with a 4-meter diameter mirror, flying in formation and feeding a "beam-combining" spacecraft (lower right).

too hot, not too cold, just a nice, warm, wet world with a comfortable dose of sunlight. If this is the case, life might be very rare indeed. But the discovery of undersea volcanic vents on the deep ocean floor in 1977 began to change our thinking. Life of a form we never knew existed thrives in superheated water, cut off from oxygen and carbon dioxide, and in total darkness. These single-celled bacteria belong to a newly-identified domain of life called Archaea, and may be the most abundant living things on Earth—the rocks below our feet may be teeming with them (see the essay by Thomas Gold in Section Five). Analysis of the genetic makeup of living species indicates that any common ancestor of all life 4 billion years ago was very possibly like Archaea—needing only liquid water, minerals, and a flow of energy (volcanic would do just fine) to flourish. If this is indeed the way life began on Earth, then life might have arisen on almost every Earth-sized planet in the Galaxy, because internal water and heat should be common in such worlds.

In our own solar system, Mars has long been the object of speculation about life beyond Earth. Pictures from NASA's interplanetary probes clearly show that abundant water flowed on Mars early in its history. In 1996, scientists reported possible evidence for past life on Mars, based on the analysis of a Martian meteorite found in Antarctica. Those claims are still controversial and the subject of a lively scientific debate, but they have spurred the development of an exciting mission, the Mars Sample Return. In the next decade, NASA and space agencies of other nations will mount an ambitious robotic expedition to Mars to collect geological specimens from these now-dry river valleys and return them to Earth for analysis. Our first secure evidence for life beyond Earth may not be far away. And NASA is planning a

journey to Europa, one of Jupiter's moons, believed to have a deep ocean beneath its icy crust, kept liquid by the heat from tides raised by the giant planet (see the essay by Clark R. Chapman in Section One). If life is common throughout the cosmos, there may be abundant evidence for its origin and perhaps even continuing evolution within our own solar system.

From the births of galaxies and the enrichment of the heavy elements, to the building of stars and planets and the appearance of life on Earth and perhaps elsewhere in our Galaxy, the questions are many, the challenges enormous, but the time seems to be upon us to know that which we have so long wondered—where did we come from, and are we alone?

Astronauts working on the Space Telescope during the
first repair mission in December 1993.

Lyman Spitzer and the Space Telescope: Profile

Lyman Spitzer (1914–1997), American astronomer and father of the Space Telescope.

Developing a major scientific instrument can be a lifelong endeavor. In the case of the Hubble Space Telescope, arguably the most successful and celebrated scientific instrument ever built, it took fifty years from conception to full realization. The most important champion of the project was Lyman Spitzer.

Spitzer (1914–1997) studied astronomy at Yale, Cambridge, and Princeton University. During World War II, he worked on the development of sonar. In 1947 he succeeded Henry Norris Russell as Chair of Princeton's Astrophysical Sciences Department. During a long and productive career, Lyman Spitzer shaped three major fields of astrophysics—interstellar matter, plasmas, and the dynamics of star clusters. His work on interstellar matter began when he noticed that elliptical galaxies have only old

stars and no nebulas of gas and dust, while spiral galaxies include young stars and nebulas. He realized that stars must now be forming in spirals from the gas and dust of the nebulas. Spitzer went on to establish the field of interstellar matter as a major branch of astrophysics.

But his greatest legacy is the Space Telescope. The idea of putting a telescope in orbit, above the obscuring veil of the Earth's atmosphere, had been suggested as far back as 1923 by the German rocket pioneer Hermann Oberth. But rockets in those days were feeble and uncontrollable devices. Then, during World War II, the Germans developed the V-2 weapon, a powerful ballistic rocket which left the atmosphere in order to come down on a distant target. It opened the way for the rocket-powered transport of payloads into space (including instruments and telescopes instead of bombs). Spitzer recalled in a discussion about the Space Telescope that the V-2 rockets "made it all seem possible."

In 1946, Spitzer wrote a report for the RAND Corporation on "Astronomical Advantages of an Extra-Terrestrial Observatory," in which he explored the advantages of a space-based telescope of five to fifteen meters in diameter. At the time of this astonishing proposal, the Palomar 5-meter telescope, then the largest in the world, was still under construction. And it was on an accessible mountaintop, not in orbital space. In his report, Spitzer said that the best reason to build a space-based telescope would be to "uncover new phenomena not yet imagined, and perhaps to modify profoundly our basic concepts of space and time."

Spitzer described two major advantages of a space telescope. First, the atmosphere absorbs almost all the ultraviolet and most of the infrared light from space before it reaches the ground. Observations at such wavelengths must necessarily be made above the atmosphere. Second, the turbulence of the atmosphere blurs the images of celestial bodies. For example, the Palomar 5-meter telescope could in principal resolve images as sharp as 0.02 seconds of arc (like a penny seen from 120 miles away), a limit determined only by the diameter of the telescope. But in reality, atmospheric turbulence limits the sharpness of these images to about 1 second of arc. For comparison, the human eye can resolve images as small as 60 seconds of arc (1/60th of a degree). So the resolution of the Palomar telescope is only 60 times better than the human eye, but if placed above the atmosphere, it would be 3,000 times better. If astronomers could put a telescope in orbit, they would finally realize the full power of the telescopes they had been using for hundreds of

years. Images of such clarity and precision had never been obtained.

In 1957, the Russians launched the first artificial Earth satellite, which stimulated the creation of the National Aeronautics and Space Administration the following year to promote the exploration of space. NASA began experiments with Orbiting Astronomical Observatories (OAOs), small satellites that would make limited observations. But the Apollo lunar landing project absorbed most of NASA's resources and orbiting telescopes were given low priority and had little success.

After the successful lunar landing program, NASA began a modest study project for a Large Space Telescope (or LST, fondly referred to as Lyman Spitzer's Telescope). In 1977 Spitzer and the astronomer John Bahcall took up the challenge of winning the necessary political support for the project. They traveled the country, convincing first their colleagues, then congressmen, and NASA headquarters, that a space telescope would be one of the greatest triumphs of astronomy.

Setbacks abounded, but their lobbying efforts slowly built up the necessary support. Once the project began in earnest and actual hardware was built, the payoff was enormous. Not only did the telescope make the most precise observations of the universe ever recorded, but technological advances tied to the project resulted in the improvement of commercial digital cameras and video recorders.

The Space Telescope was finally launched in April 1990, after a long delay following the explosion of the Space Shuttle Challenger in 1986. But when the first images were transmitted to the ground, there was a terrible disappointment. The images were blurred and showed the telltale signs of a failure in optical design. The 2.4–meter primary mirror had been crafted with exquisite accuracy but to the wrong shape! Fortunately the telescope had been designed for service visits by astronauts, and Spitzer contributed a level of optimism after this apparent disaster. He said calmly, "These things take time. We'll just have to fix it." The next Shuttle trip to visit the telescope carried a set of corrective optics which, when installed, allowed the Space Telescope to fully realize its goals.

Since then, the Hubble Space Telescope, named in honor of Edwin Hubble, discoverer of the extragalactic universe, has made impressive discoveries in many branches of astronomy. It has contributed to our understanding of the large-scale structure of the universe, and of stellar evolution and the dynamics of galaxies.

It has produced firm evidence for black holes, and provided images of objects and phenomena not even imagined in the 1940s. Lyman Spitzer was right when he predicted that the Space Telescope would profoundly affect our conception of the universe. A gentle and graceful man, loved by all who knew him, Spitzer lived to see the marvelous images produced by the fully operational Space Telescope that was his brainchild.

For the latest images from the Hubble Space Telescope, see the Space Telescope Science Institute Web site at www.stsci.edu.

216 | 217

The Eagle Nebula, as revealed by the Hubble Space Telescope. Ultraviolet light from nearby stars is evaporating and eroding these vast clouds of interstellar gas and dust, which are several light-years across. Meanwhile, gravity is forming new stars within the denser parts of the nebula.

SETI: The Next Forty Years

Jill C. Tarter

SETI–the Search for Extraterrestrial Intelligence–is now forty years old. It was born with a paper published in 1959 by the physicists Giuseppe Cocconi and Philip Morrison, who suggested that sensitive radio telescopes could be used to search for signals of intelligent origin from space. Simultaneously and independently, the astronomer Frank Drake constructed the receivers for the first active search program, which he carried out in 1960 using the 85-foot radio telescope at Green Bank, West Virginia.

Jill C. Tarter holds the Bernard M. Oliver Chair for SETI (Search for Extraterrestrial Intelligence) and is Director for SETI Research at the SETI Institute in Mountain View, California.

Figure 1: One Hectare Telescope, a coordinated array of 500 to 1,000 dish antennas, each five meters in diameter.

No verifiable evidence of an extraterrestrial signal has yet been detected. In the past four decades, eighty-five different search projects have published their negative results. But after an even longer run of negative results, the search for extrasolar planets has only recently succeeded. In a universe with a hundred billion galaxies, each with hundreds of billions of stars, it seems unlikely that we are the only intelligent creatures, but it remains a possibility. SETI observations could disprove this possibility tomorrow, or a century from now, or never.

I am often asked how I can work on a project that might not succeed in my lifetime, or ever. My reply usually revolves around the importance of the question we are trying to answer— whether there is other intelligent life in the universe. Naturally, everyone working on the project would love to find an ETI signal. But all of us realize the enormous size of the cosmic haystack we are exploring. So I don't wake up every morning expecting to find a signal that day. Instead, on awakening, I think that I may learn something new of scientific interest and help solve another technical problem to improve the search. If you don't enjoy the process of getting there, you won't do SETI for very long.

Given the vastness of this undertaking, and the exponential increase over time in our capacity to search, it's fair to say that in the past forty years we have hardly begun to explore the possibilities, and we should not be surprised that we have not yet succeeded. There is no better summary of the situation than the conclusion of the original 1959 paper: "The probability of success is difficult to estimate; but if we never search, the chance of success is zero."

In the early 1990s, NASA planned to search for artificial extraterrestrial signals in the por- tion of the microwave spectrum that pene- trates the Earth's atmosphere. That "window"

of radio frequencies extends from about 1 to 10 GHz (1 GigaHertz = 1,000 MegaHertz). The NASA researchers were looking for "technological signals" from space, which could be distinguished from natural sources of cosmic radio noise by having a degree of compression in frequency and/or time never observed from natural astrophysical objects.

Natural emissions of radio noise are always spread over a range of adjacent frequencies in the radio spectrum. In contrast, we expect that intelligent signals intended to attract our attention would occupy only a very narrow frequency band. Such a signal could transmit information across space with minimal energy and it would look utterly artificial. However, because we have no advance knowledge of the radio frequency of an intelligent signal, we must search all the billions of discernable frequencies over a large part of the radio spectrum. Searches for such signals require very large radio spectrum analyzers with extremely fine resolution, and pattern detectors to search through the data in real time. NASA adopted two search strategies: "sky surveys" looked for intrinsically strong and continuous signals coming from any direction in the sky, while "targeted searches" focused on known solar-type stars, to look for fainter signals that might be varying in time.

The sky surveys used 34-meter radio telescopes that were part of NASA's Deep Space Network (DSN), while the targeted searches used the larger telescopes belonging to the international scientific community. The High Resolution Microwave Survey, as it was called, began on October 12, 1992 but was terminated by Congress on October 1, 1993. Since that time, all SETI activity has been privately funded. Renamed Project Phoenix, to celebrate its rise from the ashes, the targeted search has continued at the non-profit SETI

Current SETI Programs

Program Name	Telescope	Frequency Coverage	% of Sky or Targets	Sensitivity
	radio	*MegaHerz*		*Watts/m^{2} **
Phoenix	Parkes/Mopra	1200–3000	209 stars	2×10^{-25}
	140' Woodbury	1200–3000	195 stars	3×10^{-25}
	Arecibo/Lovell	1200–3000	600 stars approx.	8×10^{-27}
SERENDIP IV	Arecibo	1370–1470	28% of sky, random	1×10^{-24}
SETI@home	Arecibo	1418.7–1421.2	28% of sky, random	1×10^{-25}
BETA	Harvard 26 m	1420–1740	68% of sky	1×10^{-22}
META II	Argentina 30 m	1415.4–1425.4	78 stars, 70% of sky	5×10^{-23}
SETI Australia	Parkes–multibeam random	1410–1430	80% of sky, random	
SETI League-Argus	79 personal stations	about 1420, varies	79 x 1° strips on sky	varies
	optical	*nanometers*		*photon/ns*
Columbus OSETI	Meade 10"	550		
Harvard OSETI	Oak Ridge 61"	550	2500 stars	10
Berkeley OSETI	Leuschner 30"	550		10

*10^{-25} watts per square meter is about the energy that would reach the Earth from a 25-watt light bulb on Neptune.

Institute. Today, Phoenix is attempting to complete the original goal of using large radio telescopes in both hemispheres to search 1,000 nearby stars over frequencies from 1 to 3 GHz. The sensitivity is sufficient to detect signals with power comparable to those emitted by strong terrestrial radars even from the 155 light-year distance of the farthest targeted star.

Continuing the sky surveys without access to the telescopes of the DSN has proven much more difficult. Today, the most systematic sky surveys are the BETA project at Harvard (funded by The Planetary Society), and the University of California/Berkeley SERENDIP IV project (funded by the SETI Institute and The Planetary Society), which piggybacks on astronomical investigations at the Arecibo Observatory, the world's largest radio telescope. The coverage for these surveys is substantially less than that planned for the terminated NASA program, but so is their budget. The Table above lists the parameters for all the sky surveys and targeted searches on telescopes today. It has two note-worthy aspects that would have been missing just a few years ago: optical searches (OSETI), and the SETI@home project, which actively involves the public.

In 1997, the SETI Institute convened a series of technical workshops to set out a roadmap for the next two decades. Three recommendations emerged: (1) continue the microwave searches by building a dedicated 100-meter class

The table at left attempts to summarize the characteristics of these diverse observing projects. Much more information about each one can be found at the following Web sites:

Phoenix	http://www.seti.org/science/Welcome.html#phoenix
SERENDIP IV	http://seti.ssl.berkeley.edu/serendip/serendip.html
SETI@home	http://setiathome.ssl.berkeley.edu/
BETA	http://mc.harvard.edu/seti/
META II	http://www.planetary.org/UPDATES/seti/META2/default.html
SETI Australia	http://seti.uws.edu.au/
SETI League - Argus	http://www.setileague.org/
Columbus OSETI	http://www.coseti.org/
Harvard OSETI	http://mc.harvard.edu/oseti/index.html
Berkeley OSETI	http://sag-www.ssl.berkeley.edu/opticalseti/

antenna constructed as an array of small parabolic dishes, (2) develop targeted search programs at optical and infrared wavelengths to look for extremely short pulses and to mine existing astrophysical data for ultranarrow features in the optical spectrum, and (3) begin development of an omnidirectional sky survey antenna having extremely broad frequency coverage and the ability to detect strong, transient signals from any direction. Each of these activities is now underway.

The physicist Charles Townes has long been a supporter of searching for optical and infrared signals, in light of current technologies. Our young technology can already produce a laser pulse with 10^{15} watts of power in a trillionth of a second. A planned Helios laser system will produce a pulse with a thousand times as much energy. Such a pulse, transmitted from a 5-meter telescope, could be detected with our current photon detection technologies using a receiving telescope of the same size, out to

distances of 1,000 light-years. The optical engineer Stuart Kingsley has been using a 10-inch backyard telescope to conduct a less sensitive search of this type for several years. UC Berkeley physicist Dan Werthimer constructed an instrument for the Leuschner 30-inch telescope and showed that two detectors must be used in coincidence to remove instrumental noise. Harvard physicist Paul Horowitz and his students built a system to piggyback on the optical searches for extrasolar planets being carried out by the Center for Astrophysics. To date, this optical search has looked at nearly 1,000 nearby solar-type stars.

Although it does only a limited amount of searching, SETI@home is a fabulously successful experiment in distributed computing. It employs the unused capacity of personal computers around the world to process a portion of the data collected by the SERENDIP IV sky survey at Arecibo. Raw data from the 2.5 MHz of spectral band surrounding the 1420.4 MHz frequency of interstellar hydrogen emission are recorded to tape. These tapes go to a server at University of California/ Berkeley, where they are broken into small data parcels. The data in these parcels are then electronically sent out to be processed by individuals running the SETI@home screen saver on their personal computers. The screen saver thoroughly searches the frequency-time domain with pattern detection software, and transmits the results back to the server prior to receiving another block of data to process. The popularity of this project has surprised even its origina- tors. As of this writing, SETI@home has enlisted over 2 million users of personal computers in some 224 countries.

More than forty years have passed since Cocconi and Morrison's paper was published, and it's time that microwave SETI had its own observing facility. To get there we will have to rely on the sort of private patronage that has served optical astronomy so well. This new facility would speed up the search in many different ways. To date, efforts have concentrated on improving the speed of the signal processors, and this will continue. We also need increased sensitivity, meaning larger antenna areas, so that current detection limits can be improved. The speed of the search would increase if we could observe many targets simultaneously, by electronically focusing on many parts on the sky at once, rather than using the single beam currently available at most radio telescopes.

For the past few years, radio astronomers have been working to devise a cost-effective way to build a telescope with one million square meters of collecting area (the Square Kilometer Array, or SKA). Such an instrument would be useful in many astronomical investigations, from observing the formation of the earliest structure

in the universe by mapping highly-redshifted hydrogen emission, to observing the physics of planetary formation. This system would also have many of the same requirements as we have outlined for a SETI telescope. The SETI Institute is now working with other institutions on designs for the SKA. We believe it can best be built as a coordinated array of many small parabolic dishes.

In July 1998, the SETI Institute began a collaboration with University of California/ Berkeley on the design and development of a radio telescope array with 10,000 square meters of collecting area, the One Hectare Telescope, or 1hT. The 1hT will be a one percent demonstration for technologies that should enable the SKA a decade from now. It will be a dedicated world class facility, whose multiple beams will allow continuous and simultaneous use for both SETI and radio astronomy. By adding the signals received from separate antennas with properly chosen time delays, the array can observe targets in different parts of the sky at the same time. Figure 1 is an artist's concept of this unusual array of many 5-meter dishes. It will be constructed economically from consumer goods. The 500 to 1,000 parabolas will come from the TV satellite-dish industry. The mounts will either be modifications from that industry or be assembled in-house from universal joints and truck axle bearings. The ultra-wide band antenna feed will be made inexpensively with printed circuit board technologies. The low noise amplifier will be mass produced as a single chip and cooled by miniature cryogenics (soon to be showing up on your desktop computer).

All the signals from the dishes will be brought back to a central processing facility before they are digitized, combined, and analyzed. This will require optical fibers and inexpensive transmitters being developed for the communications industry. The number of beams observed by the array will grow gracefully as the computational capability increases. This array cannot be built today, but the constant pressure from consumer markets should soon make the needed technologies available. While we wait for them to materialize, we are building a small array of seven dishes from existing (less capable, more expensive) components, to permit experimentation with beam formation, array control, and active removal of interference (primarily from satellites). The 1hT should become operational by 2005, growing in capability steadily thereafter. Its estimated cost is under $25 million. A targeted search of 100,000 stars would take about six years. A search of a million or more stars will require the SKA, improved in sensitivity by a factor of 100.

Searching a billion or more stars is best done as a sky survey, or as a targeted search of distant aggregates of stars, such as globular clusters or entire galaxies. Because the distances are so much larger, the signals must be intrinsically much stronger to be detectable on Earth, and they may be transient. Earth might well be one among many sequential targets illuminated by a powerful extraterrestrial beacon. However, if the signal arrives here only infrequently, we would probably miss it. Even if we did receive it, current search systems would be unable to make a credible case for its extraterrestrial technological nature.

This problem calls for another approach—to design an omni-directional sky survey system, a radio "fly's eye" observatory. Like the multi-faceted eyes of insects, the outputs of a large number of small receivers could be combined to form a radio image of the entire sky above the horizon. Initial work at Ohio State University and prototyping for the SKA in The Netherlands have indicated the requirements for such a system. The challenge here is the vast

amount of computation required to form an image of the entire sky, and the problem gets worse for higher frequencies. However, the exponential improvement in computational capacity over time suggests that this instrument can be built in stages. If computational costs continue to be the dominant costs of such a system, it could eventually be enlarged to cover the entire radio and optical spectrum.

The systems I have sketched above may take us twenty years into the future. Detection of a verified extraterrestrial signal by any one of them would change the landscape entirely. The detection of a second example would make it certain that there are many, and other searches would then proliferate with the knowledge that success is guaranteed. A deliberate and directional radio beacon from space would probably carry decipherable information. In contrast, detection of signals employed by a distant technology for their own "local" communication would leave us with only the proof of their existence. Yet that alone would change everything.

If a detection still eludes us forty years hence, what next? By then we will know whether life evolved independently elsewhere within our own solar system, and we will have a much clearer picture of the range of environments conducive to life. By then, we may have found indirect evidence for life in the vicinity of a nearby star and sent probes to study it more closely. If we learn that life itself is ubiquitous, then we will keep on searching for the intelligent variety in every way we can conceive, using whatever new physics and technology we have invented in the intervening decades. We might be conducting microwave searches from the lunar farside, to avoid radio pollution from terrestrial and orbital transmitters. We might be investigating the surfaces of extrasolar planets and their moons. We might have launched a multi-frequency observing platform towards a position far enough from the Sun (at over 500 times the Earth's distance) to use the Sun itself as a gravitational lens, focusing signals from the depths of the Galaxy.

Or, we may have concluded that the habitats for life are rare, and that our closest neighbors, if any, are very distant. If forty years from now we have established a stable, global-social structure, so that we can answer the questions "Who speaks for Earth?" and "What will we say?", then we might have the political and cultural will to dedicate significant resources to constructing our own transmitter(s) to make us detectable at very large distances. For this transmission to have any chance of being detected elsewhere, it must be a long-term endeavor, in astronomical terms. And the listening should continue as well. Even eighty years of effort is only the blink of a cosmic eye, and the question will still be of interest.

About the Rose Center
for Earth and Space

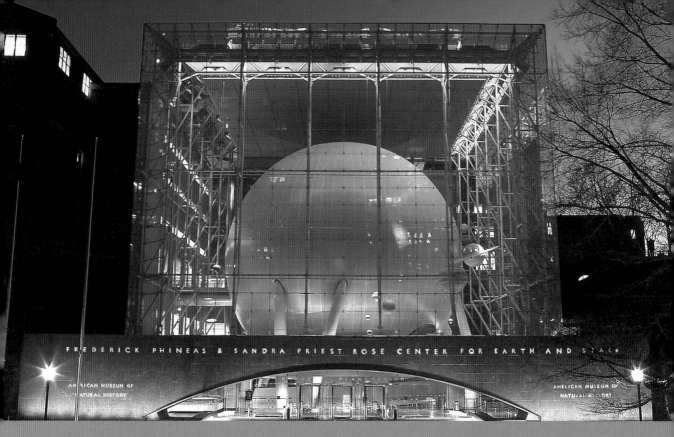

Exterior of the Rose Center for Earth and Space.

In February 2000, the American Museum of Natural History opened the Frederick Phineas and Sandra Priest Rose Center for Earth and Space. The facility serves to enhance the public's understanding of our planet and its grasp of profound astronomical concepts such as the size, age, and origin of the universe and the evolution of galaxies, stars, and planets. The Rose Center, which includes the Gottesman Hall of Planet Earth, the Cullman Hall of the Universe, and the redesigned Hayden Planetarium, links seamlessly to the rest of the Museum's permanent exhibition halls and increases the Museum's exhibition, research, and education space significantly.

The centerpiece of the new facility is an 87-foot sphere, which appears to float in space within an enormous glass cube. The upper part of the sphere contains the new Hayden Planetarium, where a powerful virtual reality simulator provides visitors with the experience of moving through a scientifically-accurate digital universe. The Planetarium also has a Zeiss Mark-9 Star Projector, which uses fiber optics to project the sky as seen from Earth with extraordinary fidelity.

The inaugural space show offers a stunning virtual voyage through the observable universe. Starting on Earth, we fly past the major planets of the solar system and then into interstellar space. As nearby stars stream by in three dimensions, we plunge into the Orion Nebula, where stars are being born. Taking leave of our Milky Way Galaxy, we then back away from the enormous Virgo Supercluster of galaxies and continue our virtual journey until we can view the large-scale structure of the universe. The production of these scenes was based on real astronomical data, in close collaboration with the astronomers who made the observations.

In the Rose Center, visitors can compare sizes of objects,
ranging from the observable universe down to a proton.

The lower half of the Hayden sphere contains the Big Bang theater. Here we follow the fiery birth of our universe, which launched the expansion of space that continues to this day. Leaving the Big Bang theater, we enter onto the Heilbrunn Cosmic Pathway. A curved ramp traces some 13 billion years of cosmic evolution, from the Big Bang to the present, at 45 million years per foot. The Rose Center also features an exhibition on the relative sizes of objects in the universe, which engages the sphere as an exhibit element. From superclusters of galaxies to subatomic particles, this exhibit extends along an elevated pathway that entirely surrounds the great sphere.

Exhibits on the main floor in the Hall of the Universe examine our modern understanding of astrophysics, arranged in four thematic zones. The Universe Zone explores the origin and evolution of the expanding universe, from the Big Bang to the formation of galaxies. The Galaxies Zone explains the architecture and dynamics of galaxies, the basic building blocks of the universe. The Stars Zone describes the many types of objects that make up the "stellar zoo," and follows the life cycles of these stars from birth to death and regeneration. The Planet Zone describes our solar system and recent discoveries of planets around other stars.

The Cullman Hall of the Universe also includes exhibits on special topics. The Supernova exhibit depicts the catastrophic explosions of certain stars, which spew heavy elements into the surrounding interstellar medium. The exhibit on Impacts features the famous Willamette Meteorite, a 15-ton iron mass from Oregon, and describes the collisions of comets and asteroids with the Earth and other targets in the solar system. The Life in the Universe exhibit features the largest sealed terrarium in

Astronomical Images

SOURCE
The National
Optical Astronomy
Observatories

WAVELENGTH

VISIBLE

2 February 2000
16:16:13 Local Time
21:16:13 Universal Time
Phase of
the Moon

The Rosette Nebula

The AstroBulletin video wall features regularly updated astronomy news.

the world. This microcosm of a planetary ecosystem lives by tapping the energy exchange between incoming sunlight and outgoing infrared radiation. The Sun exhibit, which is updated daily, features a dramatic time-lapse view of the roiling surface of our rotating star, as seen in ultraviolet light.

In all of these exhibits, the primary emphasis is on the unity of the universe and its processes. The greatest discovery of modern astrophysics is that the entire observable universe is governed by the same set of basic physical laws. The same law of gravity that draws a distant pair of galaxies together in a collision also governs the motion of a falling apple on Earth. The same laws of atomic structure that determine the colors of distant stars also govern the properties of familiar objects on Earth. The thematic exhibit clusters thus emphasize not the individual objects in the

universe, but the common principles of their origin and evolution.

Other features of the Hall include the AstroBulletin, large, high-definition video panels that display regularly-updated astronomy news, and interactive videos in which astrophysicists discuss cutting-edge questions in their fields. The Hall also includes the Black Hole theater, which describes the nature of black holes and the observational evidence for them.

The American Museum of Natural History has made a serious commitment to astrophysical research with the formation in 1999 of the Department of Astrophysics. The Department has more than a dozen Ph.D. scientists who conduct research programs in observational, theoretical, and computational astrophysics. Observational programs use both Earth-based optical telescopes and the Hubble Space

The world's largest terrarium, an allegory for a
planetary ecosystem.

Telescope. Theoretical work is focused mainly
on problems of the interstellar medium, star
formation, and stellar and galactic astrophysics.
Computational astrophysical simulations are run
at night on the same supercomputer that the
Planetarium uses during the day to generate the
digital universe programs.

Among recent initiatives, the Department of
Astrophysics received from the University of
Tokyo a GRAPE 6 (short for "gravity pipeline")
supercomputer, hard-wired to calculate the
trajectories of gravitationally interacting stars in
dense clusters. For this one specialized task,
the machine is the fastest computer in the
world. The computer, coupled by remote
access to its counterpart in Japan, is part of a
joint research project among the University of
Tokyo, the American Museum of Natural
History in New York, and the Institute for
Advanced Study in Princeton. The enormous

computational and imaging capabilities on the
Hayden Planetarium dome are being used for
research based on three-dimensional dynamical
visualizations of clusters of stars and galaxies.
This will bring to life the results of the world's
fastest computer in a way that could not be
reproduced anywhere else.

The active involvement of the Department of
Astrophysics in the development of new
planetarium shows and exhibits for the Rose
Center will assure that the American Museum
of Natural History keeps abreast of "cutting
edge" advances in the fast-moving science
of astrophysics.

Resources

Suggested Reading

Beatty, J. Kelly, Carolyn C. Petersen and Andrew Chaikin, Editors.
The New Solar System.
4th Edition. Cambridge University Press, 1998.

A first-rate survey of the solar system, with twenty-eight chapters written by experts. Up-to-date, comprehensive, and beautifully illustrated.

Ferris, Timothy.
Coming of Age in the Milky Way.
New York: Anchor Books, 1988

One of the best science writers of our time recounts the intellectual adventure of our discovery of space and time and of the physical laws that govern the universe.

Ferris, Timothy.
The Whole Shebang:
A State-of-the-Universe(s) Report.
New York: Simon & Schuster, 1997.

A summary of what we know about the cosmos and how we know it, with some speculations on the future of cosmology.

Gold, Thomas.
The Deep Hot Biosphere.
New York: Copernicus, 1999.

A revolutionary theory of a vast subterranean habitat, its significance for the origin of life on Earth, and the potential range of habitats elsewhere in the universe.

Harrison, Edward.
Cosmology: The Science of the Universe.
2nd Edition. Cambridge University Press, 2000.

A lucid exposition of cosmology, gracefully written, comprehensive, and full of historical insights and quotations.

Henbest, Nigel and Heather Couper.
The Guide to the Galaxy.
Cambridge University Press, 1994.

A guidebook to the three-dimensional "landscape" of our Milky Way Galaxy, with an historical account of its discovery.

Hoskin, Michael, ed.
The Cambridge Illustrated History of Astronomy.
Cambridge University Press, 1997.

A beautifully illustrated history of the theory and practice of astronomy from antiquity to the present day.

Kaufmann, William J. III and Roger A. Freedman.
Universe.
5th Edition. New York: W.H. Freeman, 1999.

A superb undergraduate-level textbook of astronomy and astrophysics is well written, well illustrated, and up-to-date.

Longair, Malcolm.
Our Evolving Universe.
Cambridge University Press, 1996.

A clear introduction to modern astronomy and cosmology.

Sagan, Carl.
Cosmos.
New York: Random House, 1980.

A companion piece to the PBS television series, this book gracefully examines modern astronomy in the broad context of science, culture, and history.

Sagan, Carl.
Pale Blue Dot.
New York: Random House, 1994.

A magnificently illustrated tour of our solar system.

Smolin, Lee.
The Life of the Cosmos.
New York: Oxford University Press, 1997.

A speculative thought experiment that applies Darwinian natural selection to the physical laws of the universe.

Thorne, Kip.
Black Holes & Time Warps:
Einstein's Outrageous Legacy.
New York: W.W. Norton, 1995.

The best non-technical account of the physics of black holes and the fascinating history of the subject.

Tyson, Neil de Grasse, Robert Irion, and Charles Liu.
One Universe: At Home in the Cosmos.
Washington, D.C.: Joseph Henry Press, 2000.

An exploration of the physical principles of matter, motion, and energy that govern the workings of everything from our kitchens to the expanding universe.

Verschuur, Gerrit L.
Impact! The Threat of Comets & Asteroids.
New York: Oxford University Press, 1996.

A readable account of astronomical, geological, and biological aspects of comet and asteroid impacts with the Earth.

Video

Cosmic Voyage
Smithsonian Institution and Motorola Foundation, 1996.
To order, visit
http://www.nasm.edu/nasm/pa/CV/

A dynamic 35-minute journey that shows where humans fit in the cosmic scale of things. A guided tour, from the largest to the smallest objects in the universe and of cosmic history from the Big Bang to the present. Written and directed by Bayley Silleck, narrated by Morgan Freeman.

Cosmos
Carl Sagan Productions, 2000.
To order, visit www.carlsagan.com

A guided tour of astronomy, in its cultural and historical context. Re-issued and updated version from Carl Sagan's classic 1980 PBS television series.

Web Sites

The Web sites listed below were active as of Fall 2000

American Museum of Natural History Rose Center for Earth and Space
http://www.amnh.org/rose/

Visitors can explore the themes presented in American Museum of Natural History's Rose Center for Earth and Space in this area of the Museum's Web site. The site includes the Museum's AstroBulletins, (http://astrobulletin.amnh.org/), which features the science behind current astronomical events.

Astronomical Society of the Pacific
http://www.aspsky.org/

Valuable information on professional publications, lectures, and an online catalog—a source of observing aids, books, slides, software, CD-ROMs, and other educational materials.

Astronomy Picture of the Day
http://antwrp.gsfc.nasa.gov/apod/astropix.html

The best collection of annotated astronomical images on the Internet, with text links to excellent sites. The featured image changes daily, and users can search the archives for specific images.

European Southern Observatory: Very Large Telescope Astronomical Images.
http://www.eso.org/outreach/info-events/ut1fl/astroimages.html

A collection of stunning images from the largest optical telescope array in the world.

NASA Jet Propulsion Laboratory
http://www.jpl.nasa.gov

The Jet Propulsion Laboratory (JPL) site offers information on all U.S. missions to explore the planets, including current news and images.

NASA Space Science
http://spacescience.nasa.gov/

This NASA site, focusing on the search for our cosmic origins and the structure and evolution of the universe, features astronomical images, status of missions, and educational and public outreach.

Nine Planets
http://seds.lpl.arizona.edu/nineplanets/nineplanets/nineplanets.html

This site provides an overview of the history and current scientific knowledge of each of the planets and moons in our solar system. Includes a section on current news, the search for extrasolar planetary systems, and glossary.

Space Telescope Science Institute
http://www.stsci.edu/

The Hubble Space Telescope site contains the latest pictures from HST and news releases that describe recent discoveries.

SEDS (Students for the Exploration and Development of Space)
http://www.seds.org/

An excellent student-run astronomical Web site with image catalogs and links to many useful sites.

SETI Institute Online
http://www.seti.org

This site offers information about the search for extraterrestrial intelligence and related sciences.

About the American Museum of Natural History

Founded in 1869, the American Museum of Natural History is one of the world's preeminent institutions for scientific research and education.

Today, under the direction of President Ellen V. Futter, the Museum's scientific, education, and exhibition staff are working to discover, interpret, and disseminate knowledge about human cultures, the natural world, and the universe. Prepared for the challenges of twenty-first century society, the Museum is committed not only to making contributions to science, but to improving science education and enhancing science literacy nationwide.

In 1997, the Museum launched the National Center for Science Literacy, Education, and Technology to take the Museum beyond its walls to a national audience. The National Center uses media and technology to connect people of all ages to real scientists and their work. The purpose of the National Center is to take the Museum's vast resources—collections of some 32 million specimens and artifacts, 43 exhibition halls, and a staff of more than 200 scientists, and over 130 years of expertise in educational programming—directly to classrooms, libraries, community centers, and homes throughout the country.

Contributors

Steven V. W. Beckwith is the Director of the Space Telescope Science Institute and a Professor of Physics and Astronomy at The Johns Hopkins University in Baltimore. He was educated at Cornell University and the California Institute of Technology. He has conducted pioneering research in several areas of astronomy, most notably on circumstellar disks that give rise to extra-solar planetary systems. He was a professor of astronomy at Cornell University for thirteen years and a director of the Max-Planck-Institut für Astronomie in Heidelberg, Germany for seven years before taking up his present position in Baltimore. His current research interests include the study of the birth of planets and the birth of galaxies in the early universe.

David C. Black is the Director of the Lunar & Planetary Institute and Vice President for the University Space Research Association Space Programs. Black is an Adjunct Professor in the Space Physics and Astronomy Department at Rice University. He received his Ph.D. in Physics at the University of Minnesota in 1970. He is past chairman of the Solar System Exploration Subcommittee and the Origins Subcommittee of NASA's Space Science Advisory Committee. Black has authored or co-authored over one hundred articles for scientific journals, and he was the lead editor on *Protostars and Planets II*, a fundamental resource book published by the University of Arizona Press.

Clark R. Chapman is Institute Scientist at the Department of Space Studies of the Southwest Research Institute in Boulder, Colorado. He received his Ph.D. in planetary science from Massachusetts Institute of Technology, then worked at the Planetary Science Institute in Tucson, Arizona. He was Chairman of the Division for Planetary Sciences of the American Astronomical Society. Chapman is on the imaging team of the Galileo mission to Jupiter, the imaging/spectroscopy team of the Near Earth Asteroid Rendezvous mission, and the science team of the MESSENGER mission to Mercury. His research concerns observation, interpretation, and modeling of the planets and small bodies (asteroids, comets, and planetesimals) of the solar system, and impact cratering. He has written, co-authored, and edited several professional and popular-level books.

Christopher Chyba holds the permanently endowed Carl Sagan Chair for the Study of Life in the Universe at the SETI Institute in Mountain View, California. He is also Associate Professor in the Department of Geological and Environmental Sciences and Co-Director of the Center for International Security and Cooperation at Stanford University. He chaired the Science Definition Team for NASA's Europa Orbiter mission, a mission to search for an ocean beneath the icy crust of Jupiter's moon Europa. He is also past chair of NASA's Solar System Exploration Subcommittee, which recommends priorities for solar system exploration. Chyba served on the National Security Staff of the White House from 1993 to 1995. As a consultant since leaving the White House, he drafted the President's directive on responding to emerging infectious diseases, and completed a report in 1998 on biological terrorism for the Office of Science and Technology Policy. His research centers on conditions relevant to the origin of life, and the search for life elsewhere in the solar system.

Julianne Dalcanton, a proud product of the Pittsburgh Public School system, earned her B.S. in Physics at the Massachusetts Institute of Technology. She then worked as a physicist, designing high-resolution protocols for instant magnetic resonance imaging, before resuming her education at Princeton University, where she earned her doctorate in Astrophysical Sciences while working at the campus radio station. After receiving her Ph.D., she became a Hubble Fellow at the Observatories of the Carnegie Institution in Pasadena, where she studied the formation of disk galaxies. She has recently joined the faculty of the Department of Astronomy at the University of Washington in Seattle.

Alan Dressler is an astronomer at the Observatories of the Carnegie Institution, in Pasadena, California. Dressler was educated at the University of California, receiving a B.A. in Physics in 1980 from Berkeley and a Ph.D. in Astronomy and Astrophysics from Santa Cruz in 1976. His work focuses on how galaxies are born and evolve, and on the large-scale structure of the universe and the distribution of dark matter. Working with a team of six other astronomers, Dressler found the largest structure of dark matter in the local universe, dubbed the Great Attractor. Dressler's book Voyage to the Great Attractor describes this work and puts astronomy in the larger context of human endeavor. Dressler was also one of the first to find evidence for massive black holes in the centers of nearby galaxies. A report chaired by Dressler, HST & Beyond: A Vision for Ultraviolet-Optical-Infrared Space Astronomy, urged NASA to extend the life of the Hubble Space Telescope for at least five additional years and to build of a more powerful successor, now called Next Generation Space Telescope. Both recommendations have been adopted.

Henry C. Ferguson is an Associate Astronomer at the Space Telescope Science Institute (STScI), and Deputy Project Scientist for the Next Generation Space Telescope. A 1981 graduate of Harvard College, Ferguson received his Ph.D. in Physics and Astronomy from the Johns Hopkins University. Ferguson is an author of over sixty-five research papers dealing with galaxies and galaxy evolution. He led the

STScI team that carried out the Hubble Deep Field observations with the Hubble Space Telescope in 1995 and 1998.

Adam Frank is an Assistant Professor of Astrophysics at the University of Rochester. He received his Ph.D. from the University of Washington in 1992. He has published more than 100 research papers on theoretical astrophysics. His interests center on the birth and death of stars utilizing supercomputer simulations to address issues of interstellar gas dynamics. In 1995, he was awarded the Hubble Fellowship. In 1997, he was awarded the NSF Career award. Frank is also a NASA consultant and a science writer. He has been a regular contributor to Discover and Astronomy magazines.

Thomas Gold, Professor Emeritus of Astronomy at Cornell University, is the founder and former Director of Cornell's Center for Radiophysics and Space Research. He is a Fellow of the Royal Society (London) and a Member of the National Academy of Sciences (United States). He is the author of 280 scientific papers in astronomy (including the explanation of pulsars as rotating neutron stars), cosmology, solar system research, the origin of planetary hydrocarbons, and mammalian hearing, and of the book The Deep Hot Biosphere.

Edward Harrison, Distinguished University Professor Emeritus of Astronomy at the University of Massachusetts, and member of the Five College Astronomy Department of Amherst, Hampshire, Mount Holyoke, and Smith Colleges, is Adjunct Professor at the Steward Observatory, University of Arizona. Harrison is the author of more than 200 scientific papers and several books, including Cosmology: The Science of the Universe, The Masks of the Universe, and Darkness at Night: A Riddle of the Universe.

Robert P. Kirshner is Professor of Astronomy at Harvard University and an Associate Director for Optical and Infrared Astronomy at the Harvard-Smithsonian Center for Astrophysics. A 1970 graduate of Harvard College, Kirshner received his Ph.D. in Astronomy from Caltech. Kirshner is an author of over 200 research papers dealing with supernovae, the large-scale distribution of galaxies, and the size and shape of the universe. Recent work by the High-Z Team on the acceleration of the universe was dubbed the "Science Breakthrough of the Year for 1998" by *Science* magazine. Kirshner was elected to the National Academy of Sciences in 1998.

Jane Luu was educated at Stanford and Massachusetts Institute of Technology. In 1992, she co-discovered the Kuiper Belt with David Jewitt of the University of Hawaii. Luu was an Assistant Professor at Harvard during 1994–1998, and she has since been at Leiden University in The Netherlands. Her interests include the origin of planetary systems, both in our own solar system and around other stars.

Mordecai-Mark Mac Low received his B.A. from Princeton University, and his Ph.D. from the University of Colorado at Boulder, both in physics. His research in astrophysical gas dynamics spans the range from cometary impacts on planets, through star formation in our Galaxy, to the cosmological implications of supernovae in young galaxies. He has held postdoctoral positions at NASA's Ames Research Center and the University of Chicago, and was then a scientist at the Max Planck Institute for Astronomy in Heidelberg, before joining the Department of Astrophysics at the American Museum of Natural History in 1999. One of the main tools Mac Low uses in his work is numerical simulations run on supercomputers, including the Museum's own at night, when they are not being used to run the Space Show in the Rose Center for Earth and Space.

Christopher P. McKay received his Ph.D. in AstroGeophysics from the University of Colorado in 1982 and has been a research scientist with the NASA Ames Research Center since that time. His current research focuses on the evolution of the solar system and the origin of life. He is also actively involved in planning for future Mars missions including human settlements. McKay has been involved with polar research since 1980, traveling to the Antarctic dry valleys and more recently to the Siberian and Canadian Arctic to conduct research in these Mars-like environments.

H. J. Melosh is Professor of Planetary Science at the Lunar and Planetary Lab of the University of Arizona. He received an AB degree in Physics from Princeton University in 1969, and a Ph.D. in Physics and Geology from Caltech in 1973. His principal research interests are impact cratering, planetary tectonics, and the physics of earthquakes and landslides. His recent research includes studies of the giant impact origin of the Moon, the K/T impact that extinguished the dinosaurs, the ejection of rocks from their planetary bodies and the breakup and collision of comet Shoemaker-Levy 9 with Jupiter. He is a former Guggenheim fellow (1996–1997) and was awarded the 1999 Barringer Medal of the Meteoritical Society.

David Morrison is the Director of Astrobiology and Space Research at NASA's Ames Research Center, where he manages basic and applied programs in the space, life, and Earth sciences. Internationally known for his research on small bodies in the solar system, Morrison holds a Ph.D in Astronomy from Harvard University and is the author of more than 120 technical papers and has published a dozen books. He received the Klumpke-Roberts award of the Astronomical Society of the Pacific for contributions to science education. Asteroid 2410 Morrison is named after him.

C. Robert O'Dell received his Ph.D. from the University of Wisconsin and, following a post-doctoral fellowship at the Hale Observatories, came up through the academic ranks at Berkeley and the University of Chicago. After being director of Chicago's Yerkes Observatory for six years, he joined NASA in 1972 to serve as project scientist for the Hubble Space Telescope. As the construction phase neared completion, he joined Rice University, where he had the enviable position of being a regular observer with the Hubble. He is now Distinguished Research Professor of Physics and Astronomy at Vanderbilt Unversity. His avocation is flying, with a particular passion for aerobatics. He authored a book on aerobatic flying and flew with four USA teams in world championships for glider aerobatics.

Bohdan Paczynski was educated at the Warsaw University, Poland, and he received his Ph.D. in astronomy there in 1964. He worked at the Polish Academy of Sciences in Warsaw, with frequent visits to the United States (Caltech, Princeton, Berkeley). His major work was on the evolution of single and binary stars. Since 1982 he has been a professor at Princeton. His major recent work has been in theory and observations of gravitational microlensing and theory of gamma ray bursts. He is a member of the Polish Academy of Sciences (since 1976) and the U.S. National Academy of Sciences (since 1984). His current major project is to organize the all-sky monitoring of all kinds of variable astronomical objects.

Michael M. Shara is Curator and Chair of the newly-formed Department of Astrophysics at the American Museum of Natural History. He received an M.Sc. and B.Sc. from the University of Toronto, and studied Mathematics at McGill University. He received his Ph.D. in 1977 from Tel-Aviv University. Shara then joined the Space Telescope Science Institute at Johns Hopkins,

where for seventeen years he was responsible for the peer review committees for the Hubble Space Telescope. His research interests include the structure and evolution of novae and supernovae, collisions between stars and the remnant descendants of those collisions, and the populations of stars inhabiting star clusters and galaxies. He frequently observes with the Hubble Space Telescope and other large ground-based telescopes.

Lee Smolin is a theoretical physicist whose work is mainly aimed at the discovery of the quantum theory of gravity, but he has contributed also to cosmology, astrophysics, and theoretical biology. Educated at Hampshire College and Harvard University, he has taught at Yale and Syracuse Universities and is now Professor of Physics in the Center for Gravitational Physics and Geometry at Pennsylvania State University. His first book, *The Life of the Cosmos*, is about how our understanding of the laws of physics are changed when we attempt to extend their reach to the whole universe. A New Yorker, he grew up a few blocks from the Hayden Planetarium and was first exposed to the wonders of the universe there.

Steven Soter is a member of the Department of Astrophysics at the American Museum of Natural History in New York. He studied astronomy and physics at UCLA and received a Ph.D. in astronomy in 1971 from Cornell University. He was a co-writer, with Carl Sagan and Ann Druyan, of the *Cosmos* television series. At the Smithsonian Institution's National Air and Space Museum, he was scientific curator of the gallery *How Things Fly* and worked on the development and writing of the IMAX films *Blue Planet*, *Destiny in Space*, and *Cosmic Voyage*. He was co-author, with Ann Druyan, of the Hayden Planetarium show *Passport to the Universe*. Soter is co-director of a project that located the ancient Greek city of Helike. His principle research interests are planetary astronomy and geoarchaeology.

Michael A. Strauss obtained his bachelor and doctoral degrees in Physics and Astronomy at the University of California at Berkeley. After postdoctoral positions at the California Institute of Technology and the Institute for Advanced Study, he moved to his present position at Princeton University. He has done work on the large-scale distribution and motions of galaxies, the most distant quasars in the universe, and the coolest stars known. He has been named an Alfred P. Sloan Foundation Fellow and a Cottrell Scholar of Research Corporation. In 1996, he was awarded the Newton Lacy Pierce Prize of the American Astronomical Society for his contributions to our current understanding of the large-scale structure of the universe. He currently holds the position of Associate Professor in the Department of Astrophysical Sciences at Princeton University.

Christopher Stubbs is a faculty member of both the Departments of Astronomy and Physics at the University of Washington in Seattle. His undergraduate studies were conducted at the University of Virginia, and he received a Ph.D. in Physics from the University of Washington in 1988. His research involves observational cosmology, in particular the attempt to identify the nature of the "dark matter" that pervades the universe. He is a Fellow of the Packard and Sloan Foundations, and in 1996 was presented the Award for Initiative in Research by the National Academy of Science.

Jill C. Tarter holds the Bernard M. Oliver Chair for SETI (Search for Extraterrestrial Intelligence) and is Director for SETI Research at the SETI Institute in Mountain View, California. She received her Bachelor of Engineering Physics

Degree with Distinction from Cornell University and her Masters Degree and a Ph.D. in Astronomy from the University of California, Berkeley. In her Ph.D. thesis, she coined the term "brown dwarf" to describe objects intermediate in mass between planets and stars. She served as Project Scientist for NASA's SETI program, the High Resolution Microwave Survey, and has conducted numerous observational programs at radio observatories worldwide.

Neil de Grasse Tyson is Frederick P. Rose Director of New York City's Hayden Planetarium and a member of the Department of Astrophysics at the American Museum of Natural History. Tyson's professional research interests are varied, but they primarily address problems related to star formation models of dwarf galaxies, exploding stars, and the chemical evolution history of the Milky Way's galactic bulge. Tyson has authored five books on the universe, including *The Sky is Not the Limit: Memoirs of an Urban Astrophysicist*. He is also a monthly columnist for *Natural History* magazine, and was the Project Scientist for the Hall of the Universe in the Rose Center for Earth and Space.

Roeland P. van der Marel studied astronomy and mathematics at the University of Leiden in The Netherlands, where he received his Ph.D. for a thesis on the dynamical structure of galaxies. He held a three-year Hubble Fellowship at the Institute for Advanced Study in Princeton. He is now at the Space Telescope Science Institute in Baltimore, where he was the Institute Fellow before becoming a tenure-track astronomer. His research focuses on use of the Hubble Space Telescope to find evidence for the presence of massive black holes in the centers of galaxies.

Credits

Cover

Spiral galaxy NGC 4414, courtesy of Hubble Heritage Team (AURA/STScI/NASA).

Contents

Sagittarrius star cloud, courtesy of Hubble Heritage Team (AURA/STScI/NASA)

Cygnus Loop, courtesy of Jeff Hester (Arizona State University) and NASA.

Foreword

American Museum of Natural History—circa 1900, American Museum of Natural History.

Preface

Neptune and Triton, courtesy of NASA/JPL.

Globular Cluster M80, courtesy of the Hubble Heritage Team, AURA/STScI/NASA, F. Ferraro, ESO, M. Shara, STScI, AMNH, and collaborators.

Section One

The Saturn system, courtesy of JPL/NASA.

Solar system formation, courtesy of Alfred T. Kamajian.

Mosaic of ridged plains on Europa, courtesy of NASA/JPL/CALTECH.

Close-up of Europa's ridged plains, courtesy of NASA/JPL/CALTECH.

The Conamara Chaos region on Europa, courtesy of NASA/JPL/CALTECH.

Discovery images of 1992 QB1, courtesy of Jane Luu and David Jewitt.

Plan view of Kuiper Belt orbits, courtesy of Jane Luu and David Jewitt. Illustration adapted by Diana Kline.

Schematic of the Kuiper Belt, American Museum of Natural History. Ralph Applebaum Associates.

Upsilon Andromedae, courtesy of Till Credner and Sven Kohle, Max Planck Institute for Astronomy.

Comparison of solar and planetary spectral energy distributions, courtesy of David C. Black, Lunar Planetary Institute. Illustration adapted by Diana Kline.

Infrared spectra of Venus, Earth, and Mars, courtesy of

David C. Black, Lunar Planetary Institute. Illustration adapted by Diana Kline.

Sun wobble, courtesy of David C. Black, Lunar Planetary Institute. Illustration adapted by Diana Kline.

Extrasolar systems, courtesy of David C. Black, Lunar Planetary Institute. Illustration adapted by Diana Kline.

Barringer Crater, D. Roddy, courtesy of the Lunar Planetary Institute.

Terrestrial impact frequency, courtesy of David Morrison. Illustration adapted by Diana Kline.

Ensisheim Meteorite fall, broadsheet reproduced courtesy of Universitätsbibliothek, Basel.

Ernst Florens Friedrich Chladni, reproduced courtesy of Deutschen Staatsbibliothek, Berlin.

Section Two

The Crab Nebula, courtesy of FORS Team, 8.2-meter VLT, ESO.

The Pleiades M45, © Bill and Sally Fletcher.

Eskimo Nebula, courtesy of NASA, A. Fruchter and the ERO Team, AURA/STScI.

Orion Nebula Mosaic, courtesy of C. Robert O'Dell and S. K. Wong, Rice University, AURA/STScI/NASA.

Plot of stellar luminosity versus temperature, courtesy of C. Robert O'Dell, Vanderbilt University. Adapted by Diana Kline.

Clouds of gas and dust surrounding young stars in the Trapezium Cluster, courtesy of C. Robert O'Dell, Vanderbilt University, and NASA.

Cicumstellar ring around HR 4796A, courtesy of Brad Smith, University of Hawaii, Glenn Schneider, University of Arizona, and NASA.

Protoplanetary disk, courtesy of C. Robert O'Dell, Vanderbilt University, and NASA.

Circumstellar disks, courtesy of Mark McCaughrean, Max Planck Institute for Astronomy, C. Robert O'Dell, Vanderbilt University, and NASA.

HH 30 Jet perpendicular to a dark disk seen edge on, courtesy of Chris Burrows and John Krist, AURA/STScI, Karl Stapelfeldt and colleagues, NASA/JPL/CALTECH, and the WFPC2 Science Team.

BD+31°643, courtesy of Paul Kalas and David Jewitt, AURA/STScI.

Beta Pictoris at 0.5 micron, courtesy of Paul Kalas and AURA/STScI.

Circumstellar disk around HD 141569, Alycia Weinberger, Eric Becklin (UCLA), Glenn Schneider (University of Arizona), and NASA.

NGC 3132: The Eight Burst Nebula, courtesy of the Hubble Heritage Team, AURA/ STScI/NASA.

IC 3568, round, courtesy of Howard Bond, AURA/STScI.

NGC 6826, courtesy of Bruce Balick, University of Washington, Jason Alexander, University of Washington, Arsen Hajian, U.S. Naval Observatory, Yervant Terzian, Cornell University, Mario Perinotto, University of Florence (Italy),Patrizio Patriarchi, Arcetri Observatory (Italy), and NASA.

NGC 3918, bipolar with jets, courtesy of Howard Bond, AURA/STScI.

Hubble 5, courtesy of Bruce Balick, University of Washington, Vincent Icke, Leiden University (The Netherlands), Garrelt Mellema, Stockholm University, and NASA.

NGC 7009, courtesy of Bruce Balick, University of Washington, Jason Alexander, University of Washington, Arsen Hajian, U.S. Naval Observatory, Yervant Terzian, Cornell University, Mario Perinotto, University of Florence (Italy), Patrizio Patriarchi, Arcetri Observatory (Italy), and NASA.

NGC 5307, courtesy of Howard Bond, AURA/STScI.

Friedrich Wilhelm Bessel, engraving by Johann Wolf (after E. Mandel), *Abhandlung von Friedrich Wilhelm Bessel*, Engelmann, Rudolph, ed. Leipzig: Verlag von Wilhelm Engelmann, 1875.

Companion of Sirius, courtesy of R.B. Minton.

Collision of unequal mass stars, stills from computer simulations, courtesy of Joshua Barnes, University of Hawaii.

Cecilia Payne, 1919, courtesy of the Principal and Fellows of Newnham College, Cambridge, England.

Section Three

NGC 2207 and IC 2163, courtesy of Hubble Heritage Team, AURA/STScI/NASA.

Messier 104, the Sombrero Galaxy, from VLT ANTU and FORS1, © ESO.

Spiral galaxy NGC 1232 from VLT UT 1 and FORS1 © ESO.

NGC 2997 from VLT, © ESO.

NGC 891 © WIYN Consortium, Inc. C. Howk & B. Savage (Wisconsin); N. Sharp (NOAO); WIYN/NOAO/NSF.

Cluster of galaxies in Virgo, courtesy of AURA/NOAO/NSF.

Hubble Deep Field South, courtesy of R. Williams AURA/STScI, the HDF-S Team, and NASA.

Star formation rate, courtesy of Henry C. Ferguson, AURA/STScI. Illustration recreated by Diana Kline.

Dark matter halo surrounding a galaxy. Illustration by Patrica Abt.

Gravitational lensing, courtesy of Christopher Stubbs. Illustration adapted by Diana Kline.

Portrait of Vera Rubin, courtesy of Vera Rubin.

ESO 510-13 from VLT ANTU and FORS1, © ESO.

NGC 3359, courtesy of ARC and the Sloan Digital Sky Survey.

UGC 12709, courtesy of ARC and the Sloan Digital Sky Survey.

Pegasus dSph (Pegasus Dwarf Spheroidal Galaxy): Little Galaxy of the Local Group, courtesy of Puragra Guhathakurta and Eva Grebel, W.M. Keck Observatory.

M87 and jet, courtesy of Hubble Heritage Team, STScI/AURA/NASA.

Hubble Space Telescope, courtesy of STScI/AURA/NASA.

A dust disk in NGC 7052 as seen by the Hubble Space Telescope, courtesy of Roeland P. van der Marel, STScI, Frank C. van den Bosch, NASA.

Michell's original article on black holes, courtesy of the *Philosophical Transactions of the Royal Society of London*, vol. 74, pp. 35. 1783.

Section Four

Large scale structure of the universe, simulation by Renyue Cen and Jermiah Ostriker, Princeton; visualization by Donna Cox and Bob Patterson, NCSA.

World Picture, illustration by Diana Kline. Based on original by Steven Soter.

World Map, illustration by Diana Kline. Based on original by Steven Soter.

Abell 2218: A Galaxy Cluster Lens, courtesy of Andrew Fruchter (STScI) et al., WFPC2, HST, NASA.

The distribution of roughly 1,500 galaxies in a slice six degrees thick, courtesy of Michael A. Strauss, Princeton Univertsity, from data of de Lapparent, Geller, and Huchra.

Cosmic Microwave Background, courtesy of the COBE Project, NASA.

Computer simulations of the galaxy distribution in universe dominated by Cold Dark Matter and Hot Dark Matter, courtesy of David Weinberg and Michael A. Strauss, Princeton University.

A computer simulation of data expected from the Sloan Digital Sky Survey, courtesy of David Weinberg and Michael A. Strauss, Princeton University.

BOOMERANG image of the early universe, courtesy of the BOOMERANG Project.

Bell Labs Horn Antenna, courtesy of Lucent Technologies' Bell Laboratories.

Supernova 1994D, courtesy of High-Z Supernova Search Team, HST, NASA.

Cosmic redshift. Illustration by Diana Kline. Based on original by Steven Soter.

Slices of the universe over time, courtesy of Daniels and Daniels. Illustration adapted by Diana Kline.

Evidence of expansion and acceleration. Diagram courtesy of Adam Riess and the High-Z Supernova Search Team, Space Telescope Science Institute.

Fritz Zwicky. Courtesy of the Archives, California Institute of Technology.

Allegory of the finite celestial sphere and the infinite universe beyond. From Camille Flammarion's *L'atmosphere: Meteorologie populaire*, 1888.

Ole Roemer at his telescope, 1684, courtesy of Science Museum/Science and Society Picture Library.

Jupiter and Earth orbits. Illustration by Diana Kline.

A Forest of Stars, courtesy of Edward R. Harrison, American Journal of Physics.

Schematic of new universes being born. Illustration by Diana Kline.

Georges Lemaître, courtesy of AIP Emilio Segré Visual Archives, Dorothy Davis Locanthi Collection.

Section Five

The Horsehead Nebula, courtesy of Nigel Sharp, NOAO, NSF; © AURA.

Volcanic sulfide chimney, courtesy of Deborah S. Kelley.

Earthrise over Moon from Apollo 11, courtesy of NASA.

Viking Orbiter 1 frame 897A32, courtesy of NASA/JPL/Malin Space Science Systems.

Meandering canyon on Mars, courtesy of NASA/JPL/Malin Space Science Systems.

Bacillus infernus, courtesy of Henry Aldrich, University of Florida.

The deep drilling rig at Siljan, Sweden, courtesy of Thomas Gold.

Image of Martian crater, courtesy of H. J. Melosh, University of Arizona. NASA.

Gullies in Martian crater wall, Noachis Terra, courtesy of NASA/JPL/Malin Space Science Systems.

Martian Meteorite ALH84001, courtesy of JPL/CALTECH/NASA.

Global color views of Mars, courtesy of USGS.

Mars from Viking 2 approach, courtesy USGS/NASA.

Earth, courtesy of JPL/CALTECH/NASA.

Venus from Mariner 10, courtesy of JPL/CALTECH/NASA.

Portrait of Carl Sagan, courtesy of Andy Levin/PARADE.

Section Six

Goldstone 85-foot antenna, courtesy of NASA/JPL.

Compton Gamma Ray Observatory, courtesy of STS-37 Crew, Compton Science Support Center, NASA.

Time variability of gamma ray bursts detected by BATSE, courtesy of Bohdan Paczynski. Illustration adapted by Diana Kline.

The distribution of gamma ray bursts, courtesy of Bohdan Paczynski. Illustration adapted by Diana Kline.

Super-Kamiokande neutrino observatory, ICRR (Institute of Cosmic Ray Research), The University of Tokyo.

Supernova 1987A, Hubble Heritage Team, AURA/STScI/NASA.

Three stages in the breakup of comet Shoemaker-Levy 9, courtesy of Mordecai-Mark Mac Low, American Museum of Natural History.

Model of two interacting galaxies, courtesy of Mordecai-Mark Mac Low, American Museum of Natural History.

Models of supersonic turbulence in star-forming clouds, courtesy of Mordecai-Mark Mac Low, American Museum of Natural History.

Model of the entry of a fragment of Comet Shoemaker-Levy 9 into the atmosphere of Jupiter, courtesy of Mordecai-Mark Mac Low, American Museum of Natural History.

Model of the explosion resulting from the impact of a fragment of Comet Shoemaker-Levy 9, courtesy of Mordecai-Mark Mac Low, American Museum of Natural History.

Model of multiple supernovae in the plane of a spiral galaxy, courtesy of Mordecai-Mark Mac Low, American Museum of Natural History.

Model of the cosmic web, courtesy of Mordecai-Mark Mac Low, American Museum of Natural History.

Next Generation Space Telescope, courtesy of Alan Dressler, Observatories of the Carnegie Institution and NASA.

Terrestrial Planet Finder with beams, courtesy of Alan Dressler, Observatories of the Carnegie Institution and NASA.

Hubble Space Telescope repair mission 1993, courtesy of NASA/STScI

Lyman Spitzer Jr., courtesy of Princeton University.

The Eagle Nebula, courtesy of Jeff Hester and Paul Scowen, Arizona State University, and NASA.

Artist's conception of the One Hectare Telescope, courtesy of SETI Institute.

About the Rose Center for Earth and Space

Images of the Rose Center for Earth and Space and the Cullman Hall of the Universe, courtesy of Denis Finnin, © American Museum of Natural History.

About the American Museum of Natural History

Exterior of the American Museum of Natural History–circa 1900, American Museum of Natural History.

Inside Back Cover

Orion Nebula simulation based on Hubble Space Telescope observations by C. Robert O'Dell (Vanderbilt University) and Zheng Wen (Rice University), modeled by Carter Emmart and Eric Wesselak (American Museum of Natural History Digital Galaxy Project), and rendered by David Nadeau and Jon Genetti (San Diego Supercomputer Center).

Glossary

aberration

The small angular displacement of a star's apparent position due to the Earth's orbital motion and the finite speed of light. The tiny displacement oscillates with a period of one year.

absolute zero

The lowest possible temperature (– 273ºC), at which all molecular motion and thermal radiation ceases.

accretion

The collection of material together to form a star, planet, or moon, usually mediated by a rotating disk.

active galaxy

A galaxy with an unusually strong output of energy, thought to be powered by a supermassive black hole in its core.

ammonia

A molecule composed of one nitrogen atom and three hydrogen atoms (NH_3).

Andromeda Galaxy

The nearest large spiral galaxy to the Milky Way, about 2.5 million light-years away. Also known as M31.

angular momentum

The momentum a body possesses due to its rotation or its motion in a curve, such as orbital motion, defined mathematically as the product of mass, speed, and distance from a central axis.

antimatter

Matter made of particles with the same mass as the corresponding particles of conventional matter, but with an opposite electrical charge.

arc

A measure of angular separation in the sky. The angle from the horizon to the zenith (directly overhead) is 90 degrees of arc. Both the Moon and the Sun span one-half degree of arc. One degree = 60 minutes and one minute = 60 seconds of arc.

archaea

The third major domain of life, after bacteria and eukaria. They are single-celled microbes without a nucleus, like bacteria, but genetically distinct from them. Many archaea thrive in high temperature environments.

asteroid

A small rocky or metallic body that orbits a star.

asteroid belt

The region of the solar system where most of the asteroids orbit. It lies between the orbits of Mars and Jupiter.

astrobiology

The study of extraterrestrial life.

astrometry

The branch of astronomy that deals with the precision measurement of positions and motions of stars.

astronomical unit (AU)

The average distance between the Earth and the Sun (149.6 million kilometers), used as a convenient "yardstick" for distances in the solar system.

astronomy

The scientific study of the universe.

astrophysics

The branch of astronomy that deals with the physics of astronomical objects and phenomena.

atmosphere

The gaseous envelope surrounding a star, planet, or satellite, and bound to it by gravity.

atom

The smallest unit of matter that retains the properties of a chemical element. An atom consists of a positively-charged nucleus surrounded by at least one electron.

ballistic trajectory

The path of a projectile determined by the force of gravity alone.

barycenter

The center of mass of any system of bodies moving under their mutual gravity.

baryonic

Ordinary matter, consisting of protons and neutrons.

Big Bang

The fiery birth of the observable universe in an explosion of space itself, which occurred at some time between 12 and 15 billion years ago. According to the prevailing theory, the Big Bang launched the observed expansion of the universe that continues to this day.

binary star system

Two stars orbiting their common center of gravity.

biochemistry

The chemical processes of living organisms; the scientific study of those processes.

biosphere

The part of a planet or moon (its atmosphere, waters, soil, and crustal rock) in which living organisms exist.

bipolar

Having or related to two poles.

black hole

A region in space where gravity is so strong that space closes back on itself, allowing nothing, not even light, to escape.

blueshift

The shortening of light waves due to the motion through space of a light source and an observer toward each other (the Doppler effect). It is so named because blue light is at the short-wavelength end of the visible spectrum.

blue straggler

A star that forms in a globular cluster from the collision or merger of two stars. It is hotter and bluer than its non-coalesced counterparts.

brown dwarf

An astronomical object with mass in the range between a planet and a star (greater than 1.3 percent and less than 8 percent the mass of the Sun). Brown dwarfs have a brief phase of weak nuclear fusion of deuterium (heavy hydrogen), but never become hot enough to fuse hydrogen, as do stars.

carbon

An element with six protons, formed during the red giant phase of stars. Carbon is the chemical basis of all terrestrial life.

carbon dioxide

A molecule consisting of one carbon atom and two oxygen atoms (CO_2). Its gas is denser than air. Its solid form is called "dry ice."

carbon monoxide

A molecule consisting of one carbon and one oxygen atom (CO). One of the most common molecules in the universe.

Cartesian

One who follows the philosophy of René Descartes regarding his logical analysis or interpretation of nature.

centrifugal force

The apparent force that seems to push an object moving in a curved path away from the axis of rotation or center of orbital motion.

circumstellar disk

Dust and gas forming a disk in orbit around a star. Some circumstellar disks may contain planetary systems.

comet

A small solar system body made of ice and dust that moves in an elliptical orbit around the Sun. A typical comet has a solid nucleus a few kilometers in diameter. When it nears the inner solar system, ices evaporate and form an extended and diffuse atmosphere, which is blown away from the Sun by the solar wind and radiation pressure to form a prominent tail of gas and dust.

cosmic

Of or relating to the universe as a whole.

cosmic background radiation

The microwave energy observed from all directions in the sky, at an equivalent temperature of 2.7 degrees above absolute zero, and interpreted as the residual glow from the Big Bang.

cosmic horizon

The apparent edge of the observable universe, at the distance from us in all directions that light has traveled in the age of the universe.

cosmic rays

Fast-moving, high energy, subatomic particles, mainly protons, that permeate the galaxy.

cosmic time

The absolute time common to all observers in our universe who are not moving through space; time in the world map (see definition).

cosmological constant

A theoretical long-range repulsive pressure that may accelerate the expansion of the universe. Originally, a term introduced by Albert Einstein in his equations of General Relativity to allow for a model of the universe that neither expanded nor contracted.

cosmology

The astrophysical study of the universe as a whole, including its origin, evolution, structure, and dynamics.

cosmos

The universe regarded as a whole, including all matter, energy, and space.

crater

A bowl-shaped depression on a planet or moon created from above by the impact of an extraterrestrial body or from below by a volcanic eruption.

Cretaceous

The geological Period from 144 to 65 million years ago. The so-called K/T boundary between the Cretaceous and the subsequent Tertiary Period marks the extinction of the dinosaurs.

critical density

The average density of matter that would just allow the universe to expand forever.

crust
The outermost solid layer of the Earth or of similar bodies.

cryogenic
Relating to processes at extremely low temperatures or to the technology used to produce such temperatures.

crystal carbon lattice
The highly compressed and crystalline form of carbon that comprises a white dwarf star.

dark matter
Unseen matter that is detected only by its gravitational pull on visible matter. Most of the universe is evidently made of dark matter. Its nature is yet to be determined.

dark matter halo
A roughly spherical halo of dark matter that surrounds a galaxy, including the Milky Way, and extends far beyond the region of luminous stars.

deep sea vent
Seafloor vent that releases hot, mineral-rich water from fissures at a mid-ocean ridge. Deep sea vents often create sulfide chimneys that rise up from the ocean floor like stalagmites and support biological communities based on chemical energy.

density
The amount of matter in a prescribed volume of material.

detritus
Any loose matter derived directly from an older source by disintegration or erosion and transported from its place of origin.

Doppler effect
The shift of the observed wavelengths of light, relative to the emitted wavelengths, due to the motion through space of the light source relative to the observer. A similar effect occurs for sound.

dwarf galaxy
The smallest and most common kind of galaxy.

eclipse
The obscuration of one celestial body caused by its passage behind or through the shadow cast by another body.

electromagnetic spectrum
The complete array of electromagnetic radiation (light). In order of increasing wavelength (decreasing frequency and energy), the spectrum ranges from gamma rays through X-rays, ultraviolet light, visible light, infrared radiation, microwaves to radio waves.

electron
A small subatomic particle with a negative electric charge. Electrons bound to the nucleus of an atom are responsible for its chemical properties.

element
A substance composed of atoms having the same number of protons in each nucleus.

elliptical galaxy
A galaxy that appears round or elliptical and lacks a disk with spiral arms. Such galaxies have little dust and gas, and show few signs of ongoing star formation.

energy
Any attribute of matter or electromagnetic radiation that can be converted into motion. Energy can be converted among its various forms (motion, light, mass, etc.) but the total amount of energy remains constant.

escape velocity
The speed required to escape the gravitational hold of a celestial body.

Euclidean
Characterizing the "flat" geometry developed by the ancient Greek mathematician Euclid, in which parallel lines never cross and the angles of a triangle sum to 180 degrees. This is the geometry of familiar experience, in which space is not curved.

exobiology
The study of evidence relevant to life on other worlds (synonymous with astrobiology).

exponential
A process that changes at an accelerating rate, for example with a constant doubling time.

extrasolar
Located or occurring outside the solar system, usually in other star systems.

extraterrestrial
Located or originating outside the Earth and its atmosphere.

field
A distribution throughout space of some physical quantity, such as magnetism or gravity.

fluorescence
The absorption of light by a substance at some wavelength and its re-emission at one or more longer wavelengths.

fusion
See nuclear fusion.

galactic disk
The flat disk of a spiral galaxy, which includes young stars and the gas and dust clouds from which they are formed.

galaxy
A massive, gravitationally-bound assembly of stars, interstellar clouds, and dark matter.

gamma rays
Highly energetic photons, having the shortest wavelengths and the highest frequencies in the electromagnetic spectrum.

gas-giant planet
A giant planet with a massive and deep atmosphere that surrounds a relatively small rocky core.

general relativity
The modern theory of gravity, introduced by Albert Einstein in 1916; it describes gravity as the curvature or warping of space due to the presence of matter.

giant star
A highly luminous star, approaching the end of its life, with an extended, tenuous atmosphere surrounding a hot core depleted in hydrogen.

globular cluster
A dense spherical cluster of hundreds of thousands of stars bound together by gravity.

gravitational field
The distribution through space of the attractive influence of the mass of a body.

gravitational lensing
The deflection, distortion, or magnification of light from a distant source, due to its passage through the gravitational field of a foreground mass.

gravitational waves
Disturbances or ripples in the fabric of space produced by violent events in the cosmos.

gravity
The force of attraction acting between any two masses (according to Isaac Newton); the curvature of space by matter (according to Albert Einstein).

greenhouse effect
The warming of a planet that occurs when visible sunlight is absorbed by the surface and re-radiated as infrared light which is then trapped by gases in the atmosphere.

helium
A lightweight, chemically-inert element created in the Big Bang and in stars by the fusion of hydrogen.

Hubble Deep Field
An image produced by pointing the Hubble Space Telescope at one area of the sky for 150 consecutive orbits, resulting in the "deepest-ever" optical image of the universe.

Hubble diagram
A diagram that plots the redshifts or recession velocities of galaxies versus their distances or apparent magnitudes.

Hubble law
The law stating that distant galaxies are moving away from us at rates proportional to their distances.

hydrocarbon
A chemical compound containing only hydrogen and carbon atoms.

hydrogen
The lightest, simplest, and the most abundant element in the universe.

hypersonic
Related to or capable of speeds equal to or greater than five times the speed of sound.

hyperthermophiles
Single-celled bacteria and archaea that live at temperatures exceeding about 80°C. The term means "extreme heat lovers."

inflation
A modification of Big Bang theory which proposes that the infant universe went through a very brief period of extremely rapid exponential expansion. Inflation suggests that the universe is vastly larger than its observable part.

infrared
Invisible electromagnetic radiation (light) with wavelengths longer than red light and shorter than microwaves. Infrared light occupies the spectral band extending from 0.75 to about 200 micrometers.

interferometric telescope
An observatory that combines the electromagnetic waves gathered by two or more separate telescopes so that the waves cancel (interfere) when out of phase. It can identify sources separated by extremely small angles.

intergalactic
Occurring or existing in the space between the galaxies.

interplanetary
Occuring or existing in the space between planets.

interstellar
Occurring or existing in the space between the stars.

ion
An atom with a net positive or negative electric charge, due to an unequal number of protons and electrons.

ionized gas
A superheated gas partially or totally composed of ions.

kilometer (abbreviated km)
A unit of length equal to 1,000 meters, or 0.62 miles.

kiloton
The explosive energy equivalent of 1,000 tons of TNT (or 4.2 x 10^{19} ergs).

kinetic energy
The energy inherent in a body due to its motion.

Kuiper Belt
A donut-shaped region of comets in orbit beyond Neptune, assumed to be the oldest surviving remnant of the original solar nebula and the source of short-period comets.

Kuiper Belt objects (KBOs)
The comets that populate the Kuiper Belt.

law of gravitation
The law stating that any two bodies attract each other with a force that increases in proportion to their masses and decreases in proportion to the square of the distance between them (discovered by Isaac Newton).

laws of orbital motion
Three laws discovered by Johannes Kepler: (1) A planet follows an elliptical orbit with the Sun at one focus. (2) The line from a planet to the Sun sweeps out equal areas in equal times, so the planet moves faster the closer it is to the Sun. (3) The square of a planet's orbital period is proportional to the cube of its average distance from the Sun, so distant planets have longer periods than close ones. These laws apply to all bodies in gravitational orbits, not just planets.

light-year
The distance that light travels in one year (63,000 astronomical units, or 9.46 trillion kilometers), a convenient unit of measurement for interstellar distances.

Local Group
A small group of about two dozen galaxies, including its two largest members, the Milky Way and the Andromeda Galaxy.

long-period comet
A comet with an orbital period exceeding about 200 years. Such long-period comets have very elongated elliptical orbits, and can have periods of more than a million years. They originate from the Oort cloud in the outermost reaches of our solar system.

lookback time
How long ago a distant object emitted the light we now see. For example, when we see the Andromeda Galaxy, which is 2.5 million light-years away, we are looking back 2.5 million years in time.

luminosity
The power output of a star or other luminous body.

macho (Massive Compact Halo Object)
An unseen stellar or planetary body that may contribute to the dark matter in galaxies.

magellanic clouds
Two irregular satellite galaxies of the Milky Way, visible even to the naked eye in the southern skies.

magnetic field
The region of space influenced by a magnetized body. Magnetic fields, due to motion involving charged or magnetic material, are found within the universe at every scale—from the atomic to the galactic.

magnetometer
An instrument used to measure a magnetic field.

magnetosphere
The volume of space around a star or a planet in which the global magnetic field influences the motion of charged particles.

magnitude
A measure of the brightness of an astronomical object on a logarithmic scale, with increasing numbers corresponding to decreasing brightness.

main sequence
The diagonal sequence of stars on a plot of stellar luminosity versus surface temperature, ranging from very luminous hot stars to relatively cool stars of low luminosity. Stars on the main sequence are in their stable mid-life phase of hydrogen fusion. More than ninety percent of all stars are found on the main sequence.

mantle
The part of the Earth (or other rocky body) lying between the outer crust and the central core.

mean motion
The average angular speed of a body in its orbit.

megaparsec (mpc)
3.26 million light-years.

megaton
The energy equivalent of a million tons of TNT.

meteor
A bright streak of light produced by a small fragment of rock or metal that burns up as it enters the atmosphere.

meteorite
A fragment of rock or metal that has landed on the Earth from interplanetary space. Most meteorites come from the asteroids, but a few are from other planets or satellites.

meter
The principal unit of length in the metric system, equal to 3.28 feet.

methane
The simplest hydrocarbon molecule, made of one carbon and four hydrogen atoms (CH_4). Methane is the principal constituent of natural gas.

microbe (also **microorganism**)
A microscopic single-celled living organism.

micrometer (or **micron**)
A unit of length equal to one-millionth of a meter.

microwave
Electromagnetic radiation with wavelengths between about one millimeter and thirty centimeters, intermediate between infrared light and radio.

Milky Way
The faint band of light stretching across the sky, due to myriad faint stars and nebulas; the name of the spiral galaxy containing our solar system.

molecular cloud
A large interstellar cloud of gas and dust with temperatures low enough for atoms to combine as molecules. Giant molecular clouds are the main regions of star and planet formation in galaxies.

molecule
The smallest unit of matter that retains the properties of a chemical compound. It consists of two or more atoms held together by chemical bonds between the outer electrons.

momentum
The measure of the motion possessed by a body, the product of its mass and velocity.

nanometer
One-billionth of a meter.

natural selection
The mechanism of biological evolution. The differential selection of traits most suitable for survival in reproducing entities (usually living organisms) subject to mutation.

nebula (pl. **nebulae** or **nebulas**)
An immense cloud-like mass of interstellar gas and dust, generally in the spiral arms of a galaxy.

neutrino
A subatomic particle with no electric charge and little or no mass that travels at nearly the speed of light. Neutrinos are produced in large quantities by nuclear reactions in stars.

neutron
An electrically neutral subatomic particle found in the nucleus of all atoms except ordinary hydrogen.

neutron star
An extremely dense collapsed star consisting mainly of neutrons. A neutron star is what often remains after the supernova explosion of a massive star.

nuclear
Of or relating to the atomic nucleus.

nuclear fusion
The energy-producing process in stars and the source of all the heavier elements; the combining of two atomic nuclei at high temperature to form a heavier nucleus.

nuclear physics
The study of the forces, behavior, and internal structures of atomic nuclei.

nucleus (pl. **nuclei**)
The "core" of an atom, containing the atom's protons and neutrons.

observable universe
The part of the universe lying within our cosmic horizon.

observational selection
The tendency to record those objects or phenomena that are most readily observed with the available tools or techniques, and to overlook the others.

Olbers' paradox
The puzzle of why the sky is dark at night in an infinite universe of stars. The paradox, attributed to Heinrich Olbers in 1823, was first recognized by Johannes Kepler in 1610. Solutions to the paradox invoke the cosmic horizon, the expanding universe, and the fact that the stars do not live long enough to fill up space with light.

Oort cloud

A spherical cloud of trillions of comets extending about halfway to the nearest stars and weakly bound by the Sun's gravity. Long-period comets originate from the Oort cloud.

orbit

The path of one celestial body moving around another under the force of gravity.

orbital period

The time interval for a body to complete one orbit around another.

order of magnitude

A factor or "power" of ten. Two orders of magnitude is a hundred, three is a thousand, etc.

organic molecules

Molecules formed by chemical bonds mainly between carbon atoms. Life on Earth is based on organic molecules.

Orion Nebula

A large interstellar cloud of gas and dust giving birth to stars in the constellation of Orion, about 1,500 light-years away.

oxygen

An element consisting of atoms with eight protons. Two oxygen atoms combine to make molecular oxygen (O_2) and three make ozone (O_3). The Earth's atmosphere is twenty-one percent molecular oxygen.

ozone

A molecule consisting of three oxygen atoms (O_3). Ozone in the Earth's stratosphere absorbs most of the potentially lethal solar ultraviolet radiation, preventing it from reaching the surface.

parabola

A curve that describes both the cross-section of a reflector that focuses rays to a point and the orbit of a body having escape velocity.

parallax

The difference in the direction of an object when viewed from two different locations. If the object is a star and the two locations are at opposite ends of the Earth's orbit, the measured angle of parallax allows one to calculate the distance to the star.

Pauli exclusion principle

A basic law of quantum mechanics named for Wolfgang Pauli stating that no two electrons may occupy the same small volume of space simultaneously. This law accounts for the properties of the chemical elements and the pressure that supports white dwarf stars.

perturbation

A slight disturbance in the orbit of one celestial body around another caused by the gravitational attraction of a third body

photodetector

A device to detect and measure the intensity of light.

photometric

Relating to the measurement of light.

photon

A particle or wave packet of light.

photosynthesis

The chemical process by which green plants use solar energy to synthesize carbohydrates from carbon dioxide and water, releasing molecular oxygen as a byproduct.

physics

The study of matter and energy, and the forces and fields by which they interact in space and time.

planet

An astronomical body with enough mass for its gravity to make it spherical but not enough to generate nuclear energy. Planets have non-intersecting orbits around stars or drift freely in space.

planetary nebula

An expanding shell of luminous gas that surrounds a small white dwarf star. The ionized shell receives ultraviolet light from the hot white dwarf and re-emits it as colorful visible light by fluorescence. (Planetary nebula have nothing to do with planets, the term is a historical misnomer.)

planetesimal

One of the family of asteroid-sized bodies that first condensed out of the disk of the solar nebula and later collided to form the planets.

plasma

A hot gas consisting of ionized atoms and free electrons. A fourth state of matter (as distinct from solids, liquids, and normal gases) at high temperatures.

plume

An upwelling of matter moving away from a source of heat.

plutino

A subclass of Kuiper Belt objects which, like Pluto, orbit the Sun twice during every three orbits of Neptune.

proton

A positively-charged subatomic particle. Every atomic nucleus contains one or more protons.

protoplanetary disk

A disk of dust, gas, and perhaps developing planets orbiting a young star. A transitional stage between a solar nebula and a solar system.

protostar

A gravitationally contracting gas cloud in the early stage of star formation, before fusion begins in its core.

pulsar

A rapidly spinning neutron star that emits radio energy at regular intervals and is thereby observed on Earth as a pulsating radio source.

quantum mechanics

The branch of physics that describes the interaction of matter and radiation at the atomic and subatomic levels, based on the fact that energy is observed in discontinuous units or quanta rather than being continuously divisible.

quasar

The highly luminous core of a remote galaxy, thought to be powered by a supermassive black hole. Quasars look like stars on an ordinary photograph but have very different spectra.

radiation

The emission of energy by waves (including light) or particles.

radioactivity

The emission of energetic subatomic particles and/or gamma rays from the decay of unstable atomic nuclei.

radio astronomy

The study of the universe using observations of the natural radio emissions from celestial objects.

radio telescope

A radio antenna or dish-shaped reflector used to collect and detect radio emissions from astronomical sources.

radio waves

Low energy electromagnetic radiation, with long wavelengths and low frequencies.

rarefied gas

A gas with extremely low density.

red giant

A large, highly luminous and relatively cool (red) star at a late stage of its life, once it has exhausted its core hydrogen fuel.

redshift

The lengthening of light waves due to the motion through space of a light source and an observer away from each other (the Doppler red shift), or due to the expansion of space

itself (the cosmic redshift). It is so-named because red light is at the long wavelength end of the visible spectrum.

refraction

A change in direction or the bending of light rays as they pass at an angle through different transparent substances, such as glass, water, or air.

resolution

The ability of a telescope to distinguish adjacent objects or of a spectrograph to distinguish adjacent wavelengths; the clarity of an image.

resonance

One of the natural states of oscillation in a physical system, such as the periodic swing of a pendulum or vibration of a spring.

right ascension

A measure of longitude in hours along the celestial equator.

rogue planet

A planet-sized body that escaped its host planetary system and is not gravitationally bound to a star.

satellite

A body that orbits around a larger body.

seismometer

An instrument that detects vibrations in the ground from distant earthquakes.

SETI

An acronym for the search for extraterrestrial intelligence, currently conducted with radio telescopes and optical detectors.

shock wave

An abrupt wave (with higher pressure, temperature, and density) that passes through a medium when a disturbance, such as an explosion, generates motions that exceed the speed of sound in the medium.

short-period comet

A comet with an orbital period less than about 200 years, the most famous example being Halley's Comet, which appears every seventy-six years. Short-period comets come from the Kuiper Belt and typically orbit the Sun in the same direction as the planets.

sidereal

Of or related to the stars.

solar mass

The amount of mass in the Sun, a convenient unit for expressing the mass of other stars.

solar nebula

The cloud of gas and dust that formed the young Sun and the surrounding planets.

solar system

The Sun and all the objects bound to it by gravity (planets, satellites, asteroids, comets).

spallation

The process in which small quantities of rock near the surface around a meteorite impact are ejected at high speed with only slight impact shattering.

spectroscope

An optical instrument designed to spread out light into the spectrum of its component colors.

spectrum (pl. spectra)

The range of electromagnetic radiation (light) expressed in terms of frequency or wavelength. A rainbow displays the spectrum of visible light.

spiral galaxy

A system of billions of stars with a central bulge of older stars surrounded by a flat disk with spiral arms of gas and dust nebulas and young stars. The Milky Way is a spiral galaxy.

standard candle

A class of celestial objects of known luminosity, which can be used to find the distance to associated objects. For example, a supernova of known luminosity can be used as a standard candle to calculate the distance to the galaxy in which it occurs.

star

A self-luminous body held together by gravity and with a central temperature sufficient to liberate energy by nuclear fusion.

stellar

Relating to or consisting of stars.

stellar sarallax

The minute shift in the apparent position of a star, relative to the pattern of more distant stars, when observed from opposite sides of the Earth's orbit.

stellar wind

The outflow of charged particles that a star emits into interstellar space.

strong force

The force that holds the atomic nucleus together.

subduction

The descent of a slab of crustal rock into the interior of a planet where two tectonic plates converge.

supercluster

A congregation of clusters of galaxies.

supergiant

An extremely luminous star with an extended tenuous atmosphere.

supernova (pl. supernovae or supernovas)

The catastrophic explosion of a star, which blows off most of its mass, increasing in brightness by as much as a billion times. A Type I supernova is due to the thermonuclear detonation of a compact white dwarf star which becomes unstable by accreting mass from an orbiting companion star. A Type II supernova results from the gravitational collapse of a massive star that has exhausted its nuclear fuel.

supersonic

Relating to velocities greater than the speed of sound in a given medium.

tectonic

Relating to the internal processes that deform the outer layers of a terrestrial planet or moon.

telescope

An instrument designed to gather and focus electromagnetic radiation (light) to study celestial objects and events.

terraform

The process of generating an Earth-like environment on celestial bodies that are currently inhospitable.

terrestrial

Having to do with the Earth or other rocky/metallic planets.

thermal

Relating to heat or heat transfer.

thermonuclear

Relating to the fusion of light atomic nuclei at high temperatures and the associated release of nuclear energy.

tidal friction

The dissipation of energy in a planet or satellite due to the friction from tides. The periodic deformation of the solid body and/or any oceans by tidal gravitational forces produces heat.

topography

The surface elevation of land and its variations.

trajectory

The curving path of a body in motion through space.

tsunami

A very large ocean wave that can be produced by an underwater earthquake, landslide, volcanic eruption, or the rare impact of a comet or asteroid in the ocean.

T Tauri star

A class of young star with variable luminosity, thought to be in the process of gravitational contraction before its arrival at the main sequence where it will begin to fuse hydrogen into helium.

ultracentrifuge

A laboratory device to produce centrifugal forces thousands of times stronger than gravity.

ultraviolet radiation (UV)

Invisible electromagnetic radiation (light) with wavelengths shorter than violet light and longer than X-rays. Most of the UV light from the Sun is absorbed by the Earth's atmospheric ozone layer before it reaches the surface.

universe

The physical system that encompasses all matter, energy, and space that exists (see observable universe).

vacuum

In classical physics, a region of space devoid of matter. In quantum theory, the vacuum is seething with subatomic "virtual particles" of matter and antimatter that rapidly materialize and vanish again.

velocity

The speed and direction of an object's motion.

viscous

Characterized by resistance to flow due to internal friction within a fluid.

visible light

The portion of the electromagnetic spectrum corresponding to the visible colors, with wavelengths longer than ultraviolet light and shorter than infrared radiation.

volatile

Able to vaporize at relatively low temperature. Ices of water, methane, carbon dioxide, and ammonia are volatile.

weak force

The force that mediates radioactive decay in the atomic nucleus. It is weaker than the strong force that holds the nucleus together and the force of electromagnetism, but stronger than gravity.

white dwarf

The hot, collapsed core of a red giant star after it has expelled its outer layers and ceased to produce energy by fusion. A white dwarf has a mass comparable to that of the Sun but is no larger than the Earth.

wimp (Weakly Interacting Massive Particle)

Hypothetical subatomic particle proposed as a candidate for dark matter.

world map

The universe as it actually exists at a given instant of cosmic time, with all parts depicted at the identical age. The world map represents the universe as it would appear if the speed of light were infinite. It has no cosmic horizon.

world picture

The universe as actually observed, with its more remote parts appearing younger than regions closer to the observer, due to the finite speed of light. The world picture is bounded by a cosmic horizon.

X-ray

Electromagnetic radiation with wavelengths shorter than ultraviolet light but longer than gamma rays.

X-ray astronomy

The branch of astronomy that studies celestial objects by examining the X-rays they emit.

Questions

What structures did astronomers discover as they mapped the distribution of galaxies in space? How do these structures compare in size to clusters of galaxies?

What is the significance of dark matter in attempting to explain the observed large-scale structure of the universe?

Cosmic Expansion and Acceleration

What is the evidence that the universe is expanding?

Why does a Type 1 supernova make a good "standard candle"?

What is the evidence that the expansion of the universe is accelerating?

Two Views of the Cosmos

What is the "world picture?" How does it depend on the speed of light?

What is the "world map?" Will it change with time?

What is the cosmic horizon?

What Happened Before the Big Bang?

What might a stellar mass black hole have in common with the Big Bang?

How does Smolin's hypothesis of evolving universes resemble natural selection in biology?

What observations could be used to test his hypothesis?

Section Five

Habitable Worlds

What conditions are necessary for life as we know it?

How has the concept of "habitable zones" changed with the discovery of subsurface life on Earth?

What are some potential abodes of life in our solar system and in others?

Subsurface Life on the Earth and Other Worlds

How does the biology of the volcanic deep sea vent differ from the familiar kind of biology

What limits the depth in the Earth's crust at which microbial organisms are found?

Why might subsurface life be more common that surface life in the universe?

Can Interplanetary Rocks Carry Life?

How do we know that the SNC meteorites came from Mars?

How could rocks escape from Mars without being pulverized?

Could microbes in meteorites survive the vacuum and radiation of space?

Bringing Life to Mars

Why is Mars thought to be a potential candidate for terraforming?

What would be needed increase the atmospheric pressure on Mars?

What would be needed to make a breathable Martian atmosphere?

Section Six

Gamma Ray Bursts

What are gamma ray bursts? How were they first discovered?

What technologies made it possible to identify optical counterparts to gamma ray bursts?

What processes might create explosions efficient enough to account for gamma ray bursts?

Astrophysical Computer Modeling

Why has computer modeling become increasingly used in scientific investigation?

What kinds of astrophysical processes are best investigated with computer modeling?

What are some considerations scientists must keep in mind when interpreting computational models?

NASA's Origins Program: Answering the Ancient Questions

What questions is the NASA Origins program attempting to answer?

Why is the infrared capability of the Next Generation Space Telescope essential for observing detailed spectra of very distant galaxies?

How might it be possible to find Earth-like planets around other stars?

SETI: The Next 40 Years

How does SETI resemble searching for a needle in a haystack?

What characteristic is thought to distinguish an artificial signal from the faint sources of natural radio emission from space?

What technologies are currently being used for SETI? What technologies are planned for the next twenty years?

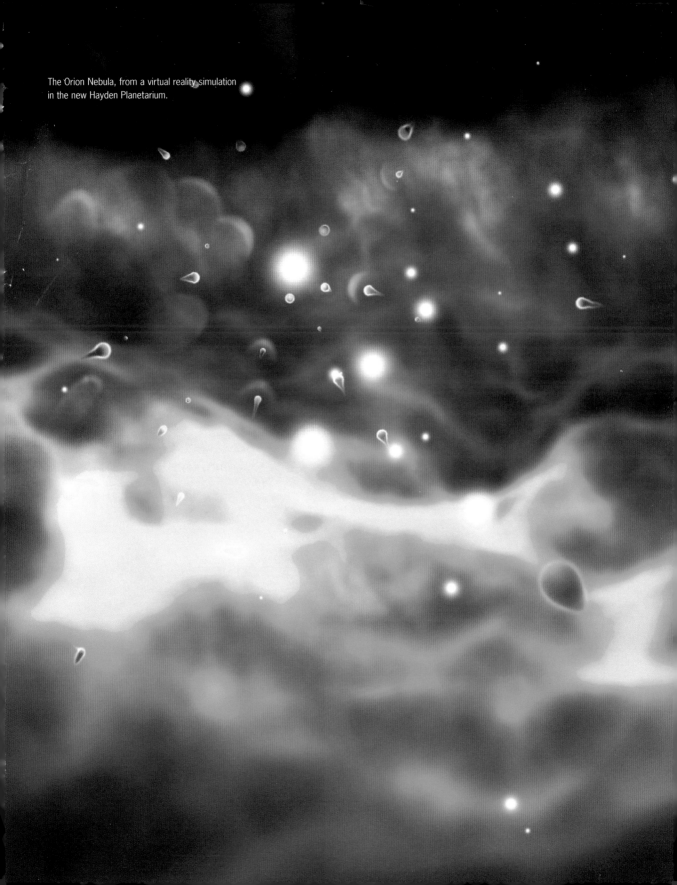

The Orion Nebula, from a virtual reality simulation
in the new Hayden Planetarium.